BULLETIN OF THE OHIO BIOLOGICAL SURVEY
New Series

Volume 10 Number 3

Larvae of the Caddisfly Genus *Oecetis* (Trichoptera: Leptoceridae) in North America

Michael A. Floyd

Published by

OHIO BIOLOGICAL SURVEY
COLLEGE OF BIOLOGICAL SCIENCES
THE OHIO STATE UNIVERSITY

Columbus, OH 43212-1192 1995

OHIO BIOLOGICAL SURVEY

Bulletin New Series: ISSN 0078-3994
Bulletin New Series Volume 10 Number 3: ISBN 0-86727-116-7
Library of Congress Number: 95-069878

CITATION

Floyd, Michael A. 1995. Larvae of the Caddisfly Genus *Oecetis* (Trichoptera: Leptoceridae) in North America. Ohio Biol. Surv. Bull. New Series Vol. 10 No. 3. viii + 85 p.

Color Separations and Film Production: Sunset Colour Productions, Ltd., Columbus, OH
Printing: The Ohio State University Printing Services
Binding: H. & H. Bookbinders, Indianapolis, IN

Ohio Biological Survey
College of Biological Sciences
The Ohio State University
Columbus, OH 43212-1192
1995

12-95—1 M

ABSTRACT

Larvae of 22 North American species of *Oecetis* are described and illustrated. Associations of larvae and adults are reported for the first time for eight species, *O. georgia* Ross, *O. morsei* Bueno-Soria, *O. nocturna* Ross, *O. osteni* Milne, *O. parva* (Banks), *O. persimilis* (Banks), *O. porteri* Ross, and *O. sphyra* Ross. Additional associations are reported for seven species which belong to the *O. inconspicua* Complex. Literature records, case descriptions, and notes on biology and distribution are given for all associated species. A key is provided for North American larvae. Notes on biology and distribution also are provided for seven, additional unassociated species, *O. daytona* Ross, *O. ditissa* Ross, *O. inconspicua* (Walker), *O. ouachita* Moulton, *O. ozarkiensis* Moulton, *O. pratelia* Denning, and *O. scala* Milne. Larval characters of *Oecetis* are discussed which support the proposed monophyly of one subgenus, *Oecetis* (*Pleurograpta*) Wallengren, and several species groups.

DEDICATION

I express my deepest appreciation to my wife, Dawn, my parents, Mr. J. Clarence Floyd and Mrs. Nancy S. Floyd, and my brother, Thomas G. Floyd. I am forever grateful for their love, patience, and encouragement and, therefore, wish to dedicate this publication to them.

ACKNOWLEDGEMENTS

I wish to express my sincere thanks to the chairman of my graduate committee, Dr. John C. Morse. I thank him for his unending patience and encouragement and for first suggesting that I work on *Oecetis* larvae. Without his direction and guidance, this project could not have been completed.

I also am grateful to my other graduate committee members, Drs. Peter H. Adler and William E. Barton, Department of Entomology, Clemson University, and Dr. J. Vaun McArthur, University of Georgia, Savannah River Ecology Laboratory, for their critical review of the manuscript and helpful suggestions.

A special expression of gratitude is extended to Dr. James B. Glover, South Carolina Department of Health and Environmental Control, and Mr. J. Kevin Moulton, University of Arizona. I appreciated their friendship, humor, and invaluable assistance on numerous collecting trips.

Additionally, I would like to thank those persons who were helpful in arranging loans of larval specimens. These persons include: Dr. Donald Azuma, The Academy of Natural Sciences of Philadelphia; Dr. Steve Brooks, The Natural History Museum, London; Drs. Phil Clausen and Ralph W. Holzenthal, University of Minnesota; Dr. David A. Etnier, University of Tennessee; Dr. Oliver S. Flint, Jr., Smithsonian Institution; Mr. David Lenat, North Carolina Division of Environmental Management; Dr. Kathleen Methven, Illinois Natural History Survey; Dr. Manuel Pescador, Florida A & M University; Mr. Arwin V. Provonsha, Purdue University; and Drs. Pat Schefter and Glenn B. Wiggins, Royal Ontario Museum.

Several others also provided specimens, shared ideas or opinions, assisted in field work, or helped in preparation of the manuscript: Mr. Douglas Bidlack, Mr. Bhodan Bilyj, Mr. Lane Bright, Dr. Y. Eric Chen, Ms. Lesa Fouts, Mr. Mike Heyn, Dr. Kevin Hoffman, Mr. John Monroe, Mr. David Scarborough, and Mr. Charles Watson. In particular, Dr. Steven Hill, Department of Biological Sciences, Clemson University, was very helpful in the identification of various species of aquatic plants.

Financial support for this research was provided under contract DE-AC09-76SROO-819 between the United States Department of Energy and the Savannah River Ecology Laboratory (SREL) Educational Program, University of Georgia. Financial support for this publication was provided by the Commonwealth Technology Corporation, Inc., and the Savannah River Ecology Laboratory (SREL) Educational Program, University of Georgia.

Michael A. Floyd, Senior Aquatic Biologist
Commonwealth Technology, Inc.
2520 Regency Road
Lexington, KY 40503-2921

CONTENTS

LIST OF FIGURES

LIST OF MAPS

INTRODUCTION

One of the largest and most diverse groups of aquatic insects is the holometabolous order Trichoptera, commonly known as caddisflies. The world fauna has been estimated at nearly 50,000 species (Schmid, 1984), and over 1,600 species have been reported from North America (Morse, 1993). Almost all descriptions, and the vast majority of systematic studies on Trichoptera in North America, have been based on the terrestrial, adult stage. The immature stages, which are found in a wide variety of aquatic habitats including small streams, large rivers, temporary ponds, reservoirs, natural lakes, bogs, and springs, have not been investigated as thoroughly. Consequently, larval descriptions are available for only about 30% of the North American species (Wiggins, 1977).

This lack of taxonomic knowledge serves to hinder progress in the discipline of benthic biology, particularly the subdiscipline of aquatic ecology. Ecological investigations of aquatic systems are based frequently on generic, rather than specific, designations of taxa. Because within a given genus there can be several species, each with unique ecological requirements, researchers cannot be sure of the actual ecological requirements of each species (Resh and Unzicker, 1975). According to Wiggins (1966), stream ecology loses a great deal due to this imbalance of knowledge, because aquatic insects, and particularly Trichoptera, constitute a large proportion of the benthic fauna upon which ecological studies are based.

This paucity of taxonomic information also hampers severely the efforts of various federal, state, and private agencies in their use of aquatic insects in monitoring stream water quality. Unfortunately, what are actually separate species must be treated as a single taxon, usually a genus, resulting in species with different biological and ecological requirements being treated equally with regard to pollution tolerance. Consequently, many researchers believe that organisms should be identified to the lowest possible taxonomic level, preferably the species (Resh and Unzicker, 1975; Lenat and Penrose, 1980; Lenat, 1988).

Although a great deal of work still remains in regard to larval systematics of Trichoptera, considerable progress has been made during the past 50 years. The classical work by Ross (1944) provided the first larval keys for the majority of Trichoptera families and genera in North America. Wiggins (1960) and Wiggins and Anderson (1968) dealt with the systematics of Phryganeidae and the limnephilid genera, *Pseudostenophylax* and *Philocasca*, respectively. Additional larval studies in the family Limnephilidae were accomplished by Flint (1960) and Wiggins (1973a; 1973b). Other investigations of larval systematics included the following: Flint (1961), *Arctopsychinae*; Flint (1962), *Rhyacophila*; Flint (1964), Psychomyiidae; Wallace and Sherberger (1970), Calamoceratidae; Sherberger and Wallace (1971), *Molanna*; Wallace (1971) and Flint (1984), *Brachycentrus*; Ross and Wallace (1974), Sericostomatidae; Chapin (1978), *Micrasema*; Schuster and Etnier (1978), *Hydropsyche* and *Ceratopsyche*; and Schefter and Wiggins (1986), *Hydropsyche morosa* group. The monumental work by Wiggins (1977), "Larvae of North American Caddisfly Genera," represented the first complete larval study of the North American genera. It provided family and generic keys, detailed illustrations of each North American genus, and discussions of biology, morphology, phylogeny, rearing, collection, and preservation. It is a must for any student of aquatic entomology and remains the standard reference for larval taxonomy in North America. A second edition of Wiggins' work is currently in progress (G. Wiggins, pers. comm.).

As for the family Leptoceridae, systematic studies of larvae and adults have been completed, or are in progress, for all eight North American genera: *Ceraclea*, *Leptocerus*, *Mystacides*, *Nectopsyche*, *Oecetis*, *Setodes*, *Triaenodes*, and *Ylodes*. Morse (1975) and Morse and Wallace (1976) revised the cosmopolitan genus *Athripsodes* and separated it into two genera, *Athripsodes* and *Ceraclea*. Resh (1976) presented larval descriptions, keys, and notes on biology for 23 *Ceraclea* species from eastern North America. A larval and adult description of the sole species of North American *Leptocerus*, *L. americanus* (Walker), was included in the classic work by Ross (1944). A comparative study of the three North American species in the genus *Mystacides* was provided by Yamamoto and Wiggins (1964). The genus *Leptocella* was synonymized with the genus *Nectopsyche* by Flint (1974). Haddock (1977) followed by discussing the biosystematics of 12 species in the genus *Nectopsyche*. The larva of an additional species, *N. tavara* (Ross), was described by Daigle and Haddock (1981). A study of the adult systematics of the North American species of *Setodes* was provided by Holzenthal (1982). The larvae of the North American species of *Setodes* currently are being studied by Miss Victoria Nations at the University of Alabama. The genus *Ylodes* was resurrected from *Triaenodes* by Schmid (1980). Larvae of the four North American species of *Ylodes* and 19 of the 23 species of

Triaenodes in North America were described by Glover (1993). Mr. Ken Manuel, Auburn University, currently is studying the systematics of the adults of the North American species of *Triaenodes*.

The adult systematics of 20 North American species of the genus *Oecetis* were included in a taxonomic revision of the *Oecetis* of the world by Chen (1993). However, *Oecetis* still remains as the only North American genus in the family Leptoceridae lacking a comprehensive study of its larval forms. In America north of Mexico, there are approximately 20 species (Wiggins, 1977), many of which are distributed widely and are often abundant. Ross (1944) remarked, "In this genus we have possibly the most widely distributed caddisflies in North America and species which become abundant in very diverse situations." Even though they often can be very abundant and common in aquatic environments, larval descriptions are available for only six North American species. Ross (1944) provided larval descriptions for five species: *O. avara* (Banks), *O. cinerascens* (Hagen), *O. eddlestoni* Ross, *O. inconspicua* (Walker), and *O. ochracea* (Curtis). Ross (1944) also provided descriptions for two additional species, *Oecetis* species A and *Oecetis* species B, whose actual identities were unknown. Based on comparison of extensive adult light-trap samples from the same localities, he suspected these two species to be *O. immobilis* (Hagen) and *O. osteni* Milne. The larval identity of *O. ochracea* also was described by Lepneva (1966) in a faunistic study of the Trichoptera larvae and pupae of the U.S.S.R. Smith and Lehmkuhl (1980) redescribed the larva of *O. avara* and provided the first larval description for *O. disjuncta* (Banks), principally a western species that is morphologically very similar to *O. avara*.

Like *Oecetis* larval taxonomy, very little is known either about the biology and life history of *Oecetis* larvae. Some general information on case construction, habitat preference, and diet has been provided (Betten, 1934; Ross, 1944; Lepneva, 1966; Cummins, 1973; Wiggins, 1977), but detailed work has been published for only three species: *O. avara, O. immobilis*, and *O. inconspicua* (Winterbourn, 1971; Shapas and Hilsenhoff, 1976; Richardson and Clifford, 1986). Food preferences have been studied by gut analysis in two species: *O. avara* and *O. inconspicua* (Winterbourn, 1971; Shapas and Hilsenhoff, 1976). There remains no biological information for at least 85% of the larvae of the *Oecetis* species that inhabit North America.

Consequently, the present study was undertaken in order to deal with these problems. The objectives for this research were: (1) to make larval-adult associations for all species of *Oecetis* and to provide illustrations and keys for larvae; (2) to provide notes on the biology of all species; (3) to describe the North American distribution of all species; and (4) to contribute larval data to an understanding of the phylogenetic relationships of the genus.

MATERIALS AND METHODS

Association and Rearing. One technique of larval-adult association is the metamorphotype method described by Vorhies (1909) and popularized by Milne (1938). This involves the collection and rearing of fully developed pupae that are enclosed within the pupal case along with the shed larval sclerites. A positive association of the different life stages can be made by collecting the shed larval sclerites from the rear of the case and matching these with the identifiable pharate adult (Wiggins, 1977), whose recognizable genitalia are visible beneath the pupal exuviae following apolysis. This is not a useful method within the Leptoceridae, however, because the larval sclerites usually are pushed from the case through the pupal cell opening during normal respiratory movements. Because of this, various other methods of laboratory rearing must be used in making larval-adult associations.

Several different methods of laboratory rearing of Trichoptera have been discussed in the literature. Resh (1972) anesthetized adult females of *Ceraclea*, identified them to species, induced them to lay eggs, and reared the eggs in the laboratory to late instar larvae that could then be described. Unfortunately, this procedure is very demanding as it requires much longer to complete, and one has to deal with the feeding of early instars (Wiggins, 1977). Furthermore, females usually do not extrude eggs while recovering from anesthetization (J. Glover, N. Hetrick, pers. comm.). Others have described different methods for rearing late-instar larvae (Ross, 1944; Philipson, 1953; Wiggins, 1959; Lepneva, 1964; Bjarnov and Thorup, 1970; Anderson, 1974; Glover, 1993). In these studies, several larvae collected from the same locality were examined under a dissection microscope to determine if they were of the same identity. Their identities were confirmed by comparing case construction, sclerite markings, or setal and gill arrangements. Some of the specimens were preserved in 80% ethanol, and the remainder were reared using a variety of methods.

In this study, larvae were reared by using a modified version of these methods. After collection and examination, individual *Oecetis* larvae were placed in small 240-ml (8 oz) plastic cups with the sides cut out and replaced with mesh screening. Each cup was filled with rocks, sand, root mats, or other suitable material to simulate the normal living conditions of the larva. The cup was sealed with a mesh-covered lid and placed in a clear, plastic rearing chamber. The rearing chamber was large enough to hold approximately eight rearing cups, and it was covered with a mesh-covered lid. The plastic container was filled approximately two-thirds full with dechlorinated water, and an electric air supply (Penn Plax XP 440, Twin) was added. The cups were monitored daily to observe for pupation and emergence, as well as deterioration of water quality within the plastic rearing chamber. The water was changed weekly or sooner if water conditions deteriorated. No effort was made to maintain water temperatures at a particular level. They were allowed to fluctuate with room temperature, approximately 21°C.

The larvae were fed primarily small (<3 mm) Simuliidae larvae (*Simulium vittatum* Complex) three to four times per week. The feeding process was accomplished by transferring each *Oecetis* larva from its rearing chamber to a small water-filled dish containing several blackfly larvae. To ensure that feeding occurred, the larvae were examined under a dissecting microscope. During each feeding, the larvae were given as many black fly larvae as they could consume (usually 5-6 individuals). Although some of the lentic species probably had never encountered a simuliid, the blackfly larvae were taken readily by them and all other species.

Collection of Reared Specimens. Upon emergence, the pharate adults or adults were collected with forceps or an aspirator and placed in labeled vials filled with 80% ethanol. For reasons unknown, many individuals did not complete development and emerge as adults. This was not a problem, however, because the pharate adults had fully developed genitalia from which positive associations could be made. Larval illustrations were produced with the use of an ocular grid and dissecting microscope (Wild, M-5A). Voucher specimens of newly associated species were deposited in the United States National Museum of Natural History (USNM), Washington, D.C.; the Illinois Natural History Survey (INHS), Champaign, Illinois; the Royal OntarioMuseum (ROM), Toronto; and the Clemson University Arthropod Collection (CUAC), Clemson, South Carolina (see List of Reared Specimens, p. 39).

Loans. Loans of additional larvae were requested from various museums, universities, and other institutions in North America. This provided additional specimens for comparison and illustration, as well as useful locality information for future collecting. Loans were obtained from Florida A & M University, INHS, North Carolina Division of Environmental Management, Academy of Natural Sciences of Philadelphia, Purdue University, ROM, University of Alabama, University of Minnesota, University of Tennessee, and USNM.

Collecting Techniques. Collections of larval material were made from a variety of habitats in both lentic and lotic environments, primarily in the southeastern United States. Collecting sites were chosen principally from examination of literature or museum records, but other sites, in close proximity to those in literature records, were also visited. Still other localities (*e.g.,* Carolina Bays, other natural lakes) were considered as being "unique" or possible refugia and, therefore, warranted investigation. Overnight collecting trips were made to Alabama, Arkansas, Florida, Kentucky, North Carolina, Oklahoma, and Ohio. Some of these trips lasted for up to one week, and two of these states, Alabama and Florida, were visited more than once. During travel on these over-night collecting trips, field-collected, live specimens were placed in rearing cups inside insulated coolers and sustained by battery-operated air pumps ("Mr. Bubbles," Marine Metal Products). At night, places of lodging were transformed into small laboratories, where the day's "catch" could be sorted, identified, and labeled. The battery-operated air pumps were replaced at night by alternating current electric pumps (Penn Plax XP 440, Twin).

Field techniques consisted primarily of using D-frame dip nets to sweep any available vegetation. The most productive sweep was done by barely scraping the bottom while pushing through the vegetation with the dip-net. Larvae were often found by breaking off the vegetation and searching it by hand in a water-filled, white pan. The submerged roots of riparian vegetation, commonly known as root mats, also were found to be very productive for some species. The roots were swept with a dip net or broken off and shaken in a pan. Other species could be found in riffle areas with the use of kick nets or seines (1.0 m^2). The overall key to finding *Oecetis* larvae was to look in as many different habitats as possible. If one particular locality was unsuccessful, then another was visited. Surprisingly, what appeared to be the same unproductive habitat searched in a previous lake or stream often would produce results in a different locality. During all collections, habitat notes were taken for each species as a means of providing biological information.

Species Accounts. The taxonomic classification used in this paper follows that of Chen (1993), who proposed four monophyletic subgenera for the genus

Oecetis based on adult characteristics. Except for original descriptions, literature records prior to 1961 are omitted. These were listed in Fischer's *Trichopterorum Catalogus* (1966; 1972). Specified states or provinces, adult collection dates, or other important information are listed in parentheses after corresponding citations. Abbreviations for larval collections, collectors, or other important terms under Material Examined are as follows: ANSP = Academy of Natural Sciences of Philadelphia, Co. = County, Cr. = Creek, CUAC = Clemson University Arthropod Collection, D. = District, FAM = Florida A & M University , GBW = G. B. Wiggins, INHS = Illinois Natural History Survey, JBG = J. B. Glover, JCM = J. C. Morse, L = larva(e), Lk. = Lake, MAF = M. A. Floyd, NCDEM = North Carolina Division of Environmental Management, OSF = O. S. Flint, Pk. = Park, PU = Purdue University, ROM = Royal Ontario Museum, ROMFP = Royal Ontario Museum field party, Rt. = route or highway, str. = stream, SCDHEC = South Carolina Department of Health and Environmental Control, TY = T. Yamamoto, UA = University of Alabama, UM = University of Minnesota, USNM = U. S. National Museum, UT = University of Tennessee. A "?" indicates an unknown county. Distribution records with an asterisk ("*") indicate new records for that state or province.

MORPHOLOGY

Larva. Mature *Oecetis* larvae range in length from 3.5 to 14 mm. Sclerotized portions of the head, thorax, and abdomen vary from dark brown to colorless. Remaining areas of the body are light green during life but white when preserved. The head usually is longer than wide, with various combinations and sizes of dark muscle scars (Figure 1). The parietal sclerites of the head capsule meet dorsally along the coronal suture. A lateral subocular line divides the parietals into a dorsal postgenal sclerite and ventral genal sclerite. However, sometimes the subocular line is very difficult to distinguish. The frontoclypeal apotome, or frontoclypeus, is separated from the parietals by the frontoclypeal sutures. The subocular lines, coronal suture, and frontoclypeal sutures make up the ecdysial lines. The frontoclypeus is subtriangular and wider anteriorly. Eyes, actually clusters of stemmata, are situated anterolaterally, usually surrounded by a lightly colored, periocular corona. Antennae are at least eight times as long as wide and usually extend forward to the posterior edge of the labrum. Sometimes, however, they are much longer, extending beyond the anterior margin of the labrum. Each antenna has a short, distal seta. The ventral apotome, or gula, incompletely separates the genal sclerites and ranges from dark brown to an almost colorless condition (Figure 3D). Sometimes it is bordered posterolaterally by much smaller, triangular sclerites (Figure 6C). The setae of the head are long and usually quite consistent in number and configuration. The mandibles are flat and knife-like. Each has three teeth, a basal, serrated, cutting edge and two lateral, basal setae. The labrum has a small, deep, median indentation surrounded by four primary setae modified as thick, curved, transparent spines. The dorsal surface has numerous, long, secondary setae. Unique among the Leptoceridae, the long, thin maxillary palpi extend well beyond the anterior margin of the labrum (Figure 2).

The pronotum is sclerotized, often distinctly pigmented, and divided by a medial fracture line (Figure 5B). There is a row of long setae running along the anterior and lateral margins and another row of five to six moderately long setae present posterior to the anterior row. Muscle scars are often present along the anterior and posterior margins. The mesonotum is sclerotized and subquadrate to subrectangular (Figure 19C). It is divided mesally into two plates and usually is much lighter than the pronotum. Its anterior margin has one seta at each anteromesal corner, up to 21 setae at each anterolateral corner, and a transverse row of three to four setae on the middle of each side. There is sometimes a row or patch of setae on the mesosternum (Figure 2). The metanotum is entirely membranous, with one to three posteromedial setae (sa2) and one to eighteen anterolateral setae (sa3). There is a transverse, irregular or regular row of two to ninety setae on the metasternum (Figure 2). The foretrochantin is subquadrate, with a pointed apex and one to eleven setae. Pleurites of the mesothorax and metathorax are lightly sclerotized and separated by a dark, narrow, pleural suture. A single seta is usually present on the episternum and epimeron ofthe mesopleuron, whereas the metapleuron may have as many as eight setae on the episternum and one seta on the epimeron. The legs are yellow to light brown and covered with dense setation. Golden brown spines are present ventrally on the tibiae and tarsi. Each foreleg is shorter but more robust and massive than the other legs. Its claw has a straight, basal spine. The midlegs and hindlegs are progressively longer. The tibiae of the hindlegs usually are undivided. Claws of the midlegs and hindlegs are often very long (sometimes longer than the respective tarsi) and curved and may have one basal seta.

Abdominal segment I has lateral and dorsal protuberances, or humps, but the lateral humps are situated somewhat ventrally and bordered by pigmented sclerites bearing rows of minute spines (Figure 2). Abdominal gills are single and present dorsally and ventrally on segments II-VIII. A pair of pleural gills is usually present on segment II. Abdominal segment II has two setae on its lateral margins. Abdominal segment VIII has one to two pairs of posterolateral setae. The ninth abdominal tergum often is sclerotized, with one lateral seta and up to 30 setae along its posterior margin. Each of the anal prolegs often has a lateral sclerite at its base. A row of four to five setae are present at the base of each of the anal claws. Each of the anal claws has one accessory hook.

Case. Case construction by larvae among the different species of *Oecetis* is highly variable (Figures 4A, 4B; 5D; 7F; 8D; 9D; 10F, 10G; 11F; 12E; 18D; 19E; 22G, 22H; 23F; 24E, 24F). Cases may be constructed of fine sand grains, rock pieces, or plant materials such as leaves and short segments of stems, twigs, or roots placed transversely. Some plant cases were observed that were somewhat spirally arranged as in the genus *Triaenodes*. There often is a mixture of plant and rock or sand pieces. Some cases are dorsoventrally flattened with larger ballast stones attached to the sides. There is often some tapering and curvature to the case, especially in early instars.

OECETIS (PLEUROGRAPTA) WALLENGREN 1891

Oecetis (Pleurograpta) testacea Group

Oecetis cinerascens (Hagen) (Figs. 3, 4, Map 1)

Setodes cinerascens Hagen. 1861. Syn. Neur. N. Am. p. 282 (Type locality: Washington, D. C., O. Sacken).

Ellis, 1962. Michigan distribution (Houghton Creek), 20 June-5 August.
Blickle and Morse, 1966. Maine distribution, 13 July-9 August.
Corbet, *et al.,* 1966. Quebec distribution and abundance (St. Helen's Island), 18 August-8 September.
Edwards, 1967. Tennessee distribution (Reelfoot Lake).
Unzicker, *et al.,* 1970. Arkansas distribution, May-June, September.
Edwards, 1973. Texas distribution, 12 April-20 June.
Longridge and Hilsenhoff, 1973. Wisconsin distribution, 7 June-2 September.
Hilsenhoff, 1975. Wisconsin distribution, keys to larvae.
Resh, 1975. Kentucky distribution, 11 May-19 September.
Resh, *et al.,* 1975. Kentucky distribution (Salt River), May-September.
Roy and Harper, 1975. Quebec distribution, 17-31 July.
Resh, *et al.,* 1978. Oklahoma and Texas distribution.
Etnier and Schuster, 1979. Tennessee distribution.
Lager, *et al.,* 1979. Minnesota distribution.
Masteller and Flint 1979, Ohio and Pennsylvania distribution (Lake Erie Region).
Neves, 1979. Massachusetts distribution, June to September.
Roy and Harper, 1979. Quebec distribution, 12 June-8 September.
Wimmer, 1979. Vermont distribution, 30 June-13 August.
Harris, *et al.,* 1980. North Dakota distribution, July-September.
Morse, *et al.,* 1980. South Carolina distribution (Upper Three Runs Creek), April-May; September.
Peterson and Foote, 1980. Ohio distribution (Furnace Run), 8 August.
Parker and Voshell, 1981. Virginia checklist.
Swegman, *et al.,* 1981. Pennsylvania distribution (Crawford Creek), May-August.
Harris, *et al.,* 1982. Florida distribution (Eglin Air Force Base), 8 November.
Holzenthal, *et al.,* 1982. Louisiana and Mississippi distribution, 11 April-8 October.
Unzicker, *et al.,* 1982. North and South Carolina distribution, mid-April-mid-October.
Flint and Butler, 1983. Maryland distribution.
Hamilton, *et al.,* 1983. Kansas distribution.
Harris, *et al.,* 1983. Alabama distribution (Tyro Creek watershed), September.
Huryn and Foote, 1983. Ohio distribution, 17 June-11 September.
Waltz and McCafferty, 1983a. Indiana distribution, May-September.
Gordon, 1984. Florida distribution.
Haag, *et al.,* 1984. Kentucky distribution (Salt River).
Harris, *et al.,* 1984. Alabama distribution (Cahaba River system), May-June; September.
Lake, 1984. Delaware distribution, May-September.
Steven and Hilsenhoff, 1984. Wisconsin distribution (Otter Creek), notes on biology.
Bilger, 1986. New England checklist (Massachussets and New York).
McCabe, 1986. New York distribution (Pine Bush), 5-16 June.
Sparks, *et al.,* 1986. Louisiana distribution.
Usis and MacLean, 1986. Ohio distribution (Stillfork Swamp), 6 June-25 September.
Downs and Goodwin, 1987. Connecticut distribution.
Nelson, 1987. Connecticut distribution.
Garono and MacLean, 1988. Ohio distribution (wetlands).
Bowles and Mathis, 1989. Arkansas distribution (mountainous regions) May-September.
Floyd and Schuster, 1990. Kentucky distribution (Buck Creek system), May-September.
Glover and Tarter, 1990. West Virginia distribution, July-October.
Harris and Lago, 1990. Alabama distribution, April-October.
Frazer, *et al.,* 1991. Alabama distribution (Little River drainage), June; September.

Tarter, 1990. West Virginia distribution, July-October.
Harris, *et al.,* 1991. Alabama distribution, April-October.
Bowles and Mathis, 1992. Oklahoma distribution, April-June; August-October.
Masteller and Flint, 1992. Pennsylvania checklist.
Mathis and Bowles, 1992. Missouri distribution (Ozark Mountains), July-August.
Floyd, *et al.,* 1993. South Carolina distribution (Upper Three Runs Creek), April; June-July; September-October.
Moulton, *et al.,* 1993. Texas distribution (Brazos River drainage), April-June; October.

DESCRIPTION. Larva: Length 7.0-9.0 mm. Head brown or dark yellow, with contrasting, white muscle scars on frontoclypeus and parietal sclerites (Figures 3A, 3B); eyes situated anterolaterally, surrounded by lightly-colored, periocular corona; antennae short and pigmented, reaching anteriorly to posterior edge of labrum; ventral apotome (VA) saucer-shaped, same color as head (Figure 3D). Pronotum brown, with white muscle scars (Figure 3B); row of long setae along anterior and lateral margins, transverse row of 4-5 setae posterior to anterior row on each half. Mesonotum light brown, with 1 seta near each anteromesal corner; 3-4 setae on each anterolateral margin, 2-3 setae in irregular row on middle of each sclerite. Metanotum membranous, each side with 1 Sa2 setae and 4-6 Sa3 setae. Metasternum with 1 pair of setae. Foretrochantins each with 1 dorsal and 1 apical seta. Basal seta present on each claw (Figures 3E, 3F, 3G); hind tibiae each with 2 rows of long setae; tarsi of midlegs and hindlegs longer than respective claws. Each side of abdomen with 1 dorsal and 1 ventral gill on each of segments II to VIII, 1 pair of pleural gills on segment II; lateral sclerites of segment I light yellow, each with 2 setae on ventral margin; abdominal tergite IX with 6 large setae, 4 small setae. **Case:** Length 8.0-10.0 mm. Bits of stems, roots, or twigs placed transversely (Fig. 4A), usually with angled appearance; sometimes with large, flat, quadrate pieces of vegetation (Figure 4B); often tapered but seldom curved.

MATERIAL EXAMINED. ALBERTA: Wildwood, Chip Lk., 2 L, 18 June 1962, GBW (ROM); same but @ Rt. 16, D. Barr; **FLORIDA:** Broward Co., Canal C-13, New Basin, St. Pkwy, 1 L, 1 Aug. 1960, W. Beck, Jr. (FAM); same but 16 Jan. 1961; Leon Co., Lk. Jackson, dead end of Fuller Rd., 4 L, 8 May 1978, Fink, Savage, and Fuller (FAM); same but 7 June 1977; same but 1 L, 27 June 1977; same but 17 June 1977; same but 6 Apr. 1978; Osceola Co., Cypress Lk., 1 L, 16 Mar. 1993, MAF and JBG (CUAC); Osceola Co., Lk. Tohopekaliga @ Southport Park, off Rt. 531, 15 km S Kissimmee, 2 L, May 1988, R. Butler (CUAC); same but 5 L, 16 Mar. 1993, MAF and JBG; Polk Co., Reedy Lk. @ roadside park nr. Frostproof, Lk. Reedy Blvd., 1 L, 18 Dec. 1991, MAF (CUAC); Putnam Co., Lk. George/St. John's R., boat ramp @ Georgetown, 1 L, 15 Mar. 1992, MAF (CUAC); Union Co., Lk. Butler @ Butler, 2 L, 15 Mar. 1992, MAF (CUAC); Washington Co., Lucas Lk. @ public boat ramp off Rt. 279, 2 L, 18 Mar. 1993, MAF and JBG (CUAC); **ILLINOIS:** Lake Co., Channel Lk. @ Antioch, 1 L, 27 May 1936, H. Ross (INHS); same but 16 May 1936, Ross and Mohr; E Fox Lk., 17 L, 12 June 1938, Ross and Burks (INHS); Fox Lk. @ Fox Lk., 38 L, 15 May 1936, Ross and Mohr (INHS); same but 3 L, 29 May 1936, Ross; same but 8 L, 28 May 1936; same but many L, 18 May 1938, Ross and Burks; Macon Co., Mt. Zion, Fork Lk., 11 L, 1940, G. Bennett (INHS); McHenry Co., Hatchery Pond @ Spring Grove, 2 L, 1938, Ross and Burks (INHS); McHenry Co., Fox R. @ Johnsburg, 5 L, 28 May 1936, Ross (INHS); **INDIANA:** Martin Co., Bogg's Cr., 4 mi. NW Shoals, 1 L, 8 Sept. 1978, A. Provonsha (PU); Noble Co., stream between Long Lk. and Duck Lk., Chain-O-Lakes St. Pk., 1 L, 24 May 1973, A. Provonsha and K. Black (PU); Stub Co., Fish Cr. @ Metz Rd., 1 L, 20 May 1972 (PU); **MARYLAND:** Montgomery Co., Potomac R. @ mth., Monocacy, 1 L, 23 Aug. 1975, S. Roback (ANSP); **MICHIGAN:** Livingston Co., E. S. George Reservoir, 1 L, 6 July 1964, Burt (ROM); **MINNESOTA:** Clearwater Co., Nicollet Cr., Itasca St. Pk., 1 L, 31 July 1974, GBW (ROM); Ramsey Co., Island Lk., Shoreview, 22 L, 7 May 1966, H. Newburg (UM); **NEW YORK:** Chautauqua Co., Lk. Chautauqua @ Long Pt. St. Pk., 1 L, 18 Sept. 1992, MAF (CUAC); **NORTH CAROLINA:** Beaufort Co., Tranters Cr., 1 L, 12 July 1989, D. Lenat (NCDEM); Craven Co., Catfish Lk. on Catfish Lk. Rd., of Rt 1100, 8 mi. WSW Havelock, 2 L, 22 May 1993, MAF and JBG (CUAC); Hyde Co., Lk. Mattamuskeet, off Rt. 94, 2mi. N Lake Comfort, 1 L, 23 May 1993, MAF (CUAC); Perquimans Co., Perquimans R. @ Lillin, 3 L, 11 July 1990, D. Lenat (NCDEM); same but @ Hertford, 2 L, 18 July 1985; Wake/Johnston Co., New R. , 2 L, July 1991, D. Lenat (NCDEM); Washington Co., Phelps Lk., off Rt. 1183, S side of Lake, 2 L, 23 May 1993, MAF (CUAC); **OHIO:** Putnam Co., Auglaize R., 3.5 mi. W Kalida, above jct. with Ottawa R., 1 L, 5 July 1955, S. Roback (ANSP); **ONTARIO:** Algoma D., Burnfield Lk., Obatanga Prov. Pk., 2 L, 6 June 1971, ROMFP (ROM); Bruce Co., Spry Lk., 1.4 mi. E Oliphant on Oliphant Rd., 5 L, 25 Aug. 1972, B. Smith (ROM); Cochrane D., Pitopiko R., crossing of Rt. 11, betw. Longlac and Hearst, 1 L, 25 June 1971, ROMFP (ROM); Kent Co., Rondeau Prov. Pk., 1 L, 3 June 1965, GBW and I. Smith (ROM); Leeds Co., Chaffey Lock, Lk. Opinicon,1 L, 21 June 1966, I. Smith (ROM); same but 27-30 June 1972; same but 26 May 1965; same but 20 June 1964, D. Barr; same but 3 L, 2 Sept. 1969, I. Smith and S. Smith; Muskoka D., Kimberly Isl., Moon R., Bala, 3 L, 10 June 1975, GBW and TY (ROM); Ontario Co., Pickering, 1 L, 26 Aug. 1967, Smith and Odum (ROM); same but 16 July 1966,

Smith, Tilton, and Odum; Parry Sound D., Gurd Lk., Grundy Lk. Prov. Pk., Rt. 69 S Sudburg, 10 L, 1 June 1971, ROMFP (ROM); Peel Co., Heart Lk., Brampton, 12 L, 6 Aug. 1966, I. Smith and S. Tilton (ROM); Peterborough Co., Lk. Katchiwano, 1 L, 12 July 1966, I. Smith (ROM); same but 2 L, 10 Aug. 1969, I. Smith; Peterborough Co., Nogies Cr. @ Fish Reservoir, 3 L, 14 June 1975, TY (ROM); same but 2 L, 7 May 1975; Simcoe Co., Orillia, Lk. Simcoe @ narrows, 3 L, 31 Aug. 1967, I. Smith and L. Kohalmi (ROM); Thunder Bay D., Lk. Kenagamisis @ Rt. 11, 15 L, 14 June 1966, L. Smith (ROM); same but 2 L, 23 June 1971, ROMFP; Wentworth Co., Hamilton, 1 L, Nov. 1973, D. Davies (ROM); **SASKATCHEWAN:** Lk. Madge, Duck Mtn. Prov. Pk., 3 L, 29 June 1962, GBW (ROM); Margo Lk., southern Saskatchewan, 2 L, 10 July 1938, D. Rawson (INHS); Murray Lk. nr. N Battleford, 1 L, 15 June 1962, GBW (ROM); **SOUTH CAROLINA:** Anderson Co., Cuffie Cr. @ Rt. 115, 1 L, R. Holzenthal (CUAC); Barnwell Co., PAR Pond, Savannah R. Site, 1 L, 23 June 1972, J. Richardson (ANSP); same but 1 L, 9 Apr. 1974; same but 9 L, 10 Apr. 1974; Berkeley Co., Cooper R., E Br. above Mayrantlead, nr. Richmond Plant., 1 L, 30 Aug. 1973, J. Richardson (ANSP); same but above Wadboo Cr. nr. Moncks Corner, 3 L, 31 Aug. 1973, J. Richardson (ANSP); same but Hagan Isl. nr. Cypress Gardens, 4 L, 28 Aug. 1973; Berkeley Co., Four Holes Swamp, Mims Lk., 1 L, July 1982, Brunswig (CUAC); Kershaw Co., Wateree R. nr. Camden, 1 L, 31 Aug. 1958, S. Roback (ANSP); Sumter Co., Touchberry Cr. @ GSX-Pinewood, 1 L, 20 Nov. 1990, Heyn, Bouknight, and Quattlebaum (CUAC); same but 18 L, 31 May 1991, M. Heyn; same but 1 L, 17 Sept. 1991; 1 mi. S Cane Savannah on St. Paul's Church Rd., 3 L, 2 Sept. 1986, P. Grant; **WISCONSIN:** Dane Co., Prairie Pond, Univ. Wisconsin Arboretum, 1 L, 10 Sept. 1974, C. Gorsuch (CUAC); Jefferson Co., Bark R. @ Rome, 4 L, 21 May 1938, Ross and Burks (INHS)

ASSOCIATION, BIOLOGICAL NOTES, AND DISTRIBUTION. Association of larvae and adults was based on rearings first accomplished by Ross (1944). Additional associations were made in this study from the rearing of specimens from Florida (Osceola County, Lake Tohopekaliga at Southport Park and Polk County, Reedy Lake near Frostproof) and South Carolina (Pickens County, Willards Pond, Clemson University dairy farm) (List of Reared Specimens, p. 39). Although the larva of this species is found in both lentic and lotic environments, it has been collected most abundantly in lentic situations. It is one of the few *Oecetis* species that frequently invades artificial ponds and reservoirs. Those specimens collected from streams or rivers usually are found in pools or other areas with slow current. Within either environment, the larvae are collected most frequently in and around aquatic vegetation (*e.g., Hydrilla* sp., *Myriophyllum* sp., *Panicum* sp., *Potamogeton* spp.). Consequently, their peculiar, "log-cabin" type, plant case may provide the secondary function of camouflage. Despite this camouflage, however, they can be prey for small fish (Ross, 1944). I examined 11 specimens from Fork Lake, Macon County, Illinois, that were removed from the stomach of a bluegill, *Lepomis macrochirus* Rafinesque. The larvae have the ability to swim due to the oar-like function of the long setal fringes of the hind tibiae. This swimming behavior was observed many times during the rearing process. Ross (1944) indicated that *O. cinerascens* was frequently taken in great numbers in Illinois, displaying a constant cycle of generations (multiple cohorts). This species is widely distributed in the eastern United States and Canada (Map 1). It is known from Alberta*, Alabama, Arkansas, Connecticut, District of Columbia, Delaware, Florida, Georgia, Illinois, Indiana, Kansas, Kentucky, Louisiana, Maine, Maryland, Michigan, Minnesota, Missouri, Mississippi, New Brunswick, North Carolina, North Dakota, New Hampshire, New Jersey, Nova Scotia, New York, Ohio, Oklahoma, Ontario, Pennsylvania, Quebec, South Carolina, South Dakota, Saskatchewan, Tennessee, Texas, Virginia, Vermont, West Virginia, and Wisconsin.

DIAGNOSIS. The larva of *O. cinerascens* is morphologically distinct from other *Oecetis* species in its head coloration. Instead of the usual light yellow or brown head with darker muscle scars found on most *Oecetis* larvae, the head color of *O. cinerascens* is dark with pale muscle scars.

Oecetis georgia Ross (Figure 5, Map 2)

Oecetis georgia Ross. 1941. Trans. Amer. Entomol. Soc. 67:98 (Type locality: Georgia, Crawford Co., 5 mi. S Roberta, 12 May 1939, P. Fattig).

Morse, *et al.,* 1980. South Carolina distribution (Upper Three Runs Creek), May-June; August-September.
Harris, *et al.,* 1982. Florida distribution (Eglin Air Force Base), 19 September.
Holzenthal, *et al.,* 1982. Louisiana distribution, 6 August-17 October.
Unzicker, *et al.,* 1982. North and South Carolina distribution, mid-April-July.
Gordon, 1984. Florida distribution.
Harris and Lago, 1990. Alabama distribution, April-September.

Harris, *et al.,* 1991. Alabama distribution, April-September.
Floyd, *et al.,* 1993. South Carolina distribution (Upper Three Runs Creek), April-October.

DESCRIPTION. Larva: 5.0-6.0 mm. Head light yellow to colorless, with few, very faint muscle scars on postgenal parietals (Figure 5A); antennae reaching to middle of labrum; ventral apotome saucer-shaped and usually colorless (Figure 5C). Pronotum light yellow to colorless, row of long setae on anterior and lateral margins, transverse row of 5-6 setae on middle of each sclerite (Figure 5B). Mesonotum colorless, with 1 seta near each anteromesal corner; 4 setae on each anterolateral corner, 2-3 setae in irregular row on middle of each sclerite. Mesosternum with 1 pair of setae. Metanotum membranous, each side with 1 Sa2 seta and 3-4 Sa3 setae. Foretrochantins with 3-4 dorsal setae. Hind tibiae each with 2 rows of long setae (Figure 3G); tarsi of midlegs and hindlegs longer than respective claws; basal seta on each claw. Each side of abdomen with 1 dorsal and 1 lateral gill on each of segments II to VII, pair of pleural gills on segment II; each lateral sclerite of segment I with 1 seta on ventral margin; abdominal tergite IX with 6 large setae and 4 small setae. **Case:** Length 6.0-7.0 mm. Bits of stems, roots, or twigs placed transversely (Figure 5D), usually tapered and curved with rounded outline.

MATERIAL EXAMINED. ALABAMA: Baldwin Co., Pine Log Cr. @ Rt. 59, 4 L, 13 Mar. 1992, MAF and JBG (CUAC); Escambia Co., Burnt Corn Cr. @ Rt. 41, 2 L, 15 Dec. 1991, MAF (CUAC); **FLORIDA:** Clay Co., Black Cr., Camp Blanding, 4 L, 6 Mar. 1991, L. Burton (CUAC); Duval Co., Trib. to Six Mi. Cr. @ Rt. 17, 6 L, 15 Mar. 1993, MAF and JBG (CUAC); Strawberry Cr., 4 L, 16 July 1952, W. Beck, Jr. (FAM); Gadsden Co., Telogia Cr. @ Rt. 270A, 3 L, 1 July 1987, M. Pescador (FAM); Hamilton Co., Highlands Hammock St. Pk., str. nr. Ranger Stat., 2 L, 16 Mar. 1993, MAF and JBG (CUAC); Okaloosa Co., Blackwater R. @ Rt. 4, 4 L, 15 Dec. 1991, MAF (CUAC); Taylor Co., Fenholloway R. @ Fenholloway, 1 L, 29 May 1991, Jones, Hubbard, and Silvanima (FAM); same but 4 L, 31 July 1991, Hubbard and Pescador; Encofina R. @ Rt. 27, 1 L, 21 Aug. 1991, Hubbard, Jones, Silvanima (FAM); same but 19 June 1991; **GEORGIA:** Crawford Co., Spring Cr. above pond @ Camp Eunice, 5 mi. SSE Roberta, 1 L, 6 June 1981, Hamilton and Rothschild (CUAC); Decatur Co., Attapulgus Cr. @ Mill White Rd., 3 L, 12 June 1991, Pescador, Jones, and Silvanima (FAM); same but 1 L, 14 Aug. 1991, Hubbard, Pescador, and Silvanima; **NORTH CAROLINA:** Columbus Co., Juniper Cr., 17 L, 17 June 1991, D. Lenat (NCDEM); Hoke Co., Juniper Cr., 2 L, 23 Apr. 1990, D. Lenat (NCDEM); same but 9 L, 20 Apr. 1990; Scotland Co., Lumber R. nr. Wagram, 6 L, 14 July 1986, D. Lenat (NCDEM); **SOUTH CAROLINA:** Aiken Co., Mill Cr. @ Rd E-2, Sav. R. Site, 2 L, 8 July 1977, Herlong and Prichard (CUAC); same but 1 L, 22 July 1977; Boggy Gut Cr. @ Rd. 781-4, Sav. R. Site, 1 L, 29 Mar. 1977, Herlong and Prichard (CUAC); same but 13 June 1977; Upper Three Runs Cr., @ Rd 8-1, Sav. R. Site, 1 L, 31 Mar. 1979, Herlong and Prichard (CUAC); same but 13 June 1977; same but 8 Aug. 1977; same but 26 Feb. 1977; same but 4 Apr. 1980; same but 8 Aug. 1977; same but 3 May 1977; same but 20 Aug. 1979, Kelley and McEwan; same but 6 Apr. 1983, MAF; Upper Three Runs Cr. @ Rt. 114, 4 L, 6 Sept. 1991, MAF and J. Moulton (CUAC); same but 1 L, 11 Sept. 1991; same but 5 L, 12 Sept. 1991; Upper Three Runs Cr. @ Rt. 278, 1 L, 27 Apr. 1992, MAF (CUAC); Upper Three Runs Cr. @ Rd. C, 1 L, 5 June 1972, J. Richardson (ANSP); Edisto R., 5 L, 30 Mar. 1990, M. Heyn (CUAC); Kershaw Co., Kelley Cr. @ I-20, Hardwicke Chem. Co., 2 L, 27 May 1991, M. Heyn (CUAC); Orangeburg Co., Edisto R. nr. Orangeburg, 1 L, 14 July 1964, S. Roback (ANSP).

ASSOCIATION, BIOLOGICAL NOTES, AND DISTRIBUTION. Positive association for this species was based on rearings of larval material from Upper Three Runs Creek, Aiken County, South Carolina (List of Reared Specimens, p. 39). This strictly lotic species sometimes inhabits aquatic macrophytes but usually is found among submerged roots of riparian vegetation (root mats) or logs, sticks, and other woody debris on the stream bottom. Its woody, log-cabin style case is very well concealed in the latter two habitats. Those streams and rivers from which the larvae have been collected range in width from <1 m to 10 m. *Oecetis georgia* has the ability to "swim" by using the long setal fringes of the hind tibiae, structures shared by *O. cinerascens* and *O. persimilis* (Banks). Larvae of this southeastern endemic have been collected only in blackwater streams of the Coastal Plain and Sandhills regions. Adults have been collected most abundantly in the Coastal Plain (Harris, *et al.,* 1991; Floyd, 1993) but also have been found in small numbers at several isolated localities in the Piedmont and Mountains of Alabama (Harris, *et al.,* 1991) and North Carolina (Floyd, unpublished data). The larvae possibly coexist with those of *O. persimilis*, because adults of the two species sometimes are collected together in the same light trap. *Oecetis georgia* is known from only six states: Alabama, Florida, Georgia, Louisiana, North Carolina, and South Carolina (Map 2).

DIAGNOSIS. *Oecetis georgia* is morphologically most similar to *O. persimilis*, a species which occurs in similar habitats and builds an identical case. The two can be separated based on the presence or absence of dark muscle scars on the head. The muscle scars are absent on *O. georgia* and present on *O. persimilis*.

Oecetis persimilis (Banks) (Figure 6, Map 3)

Oecetina persimilis Banks. 1907. Proc. Entomol. Soc. Wash. 8:129 (Type locality: Maryland, High Island and Virginia, Glencarlyn, June).

Bilger, 1986. Maine distribution, 10 July-7 August.
Unzicker, *et al.,* 1970. Arkansas distribution, May-July.
Hilsenhoff, *et al.,* 1972. Wisconsin distribution (Pine-Popple River), 8-15 July.
Edwards, 1973. Texas distribution, 31 May.
Longridge and Hilsenhoff, 1973. Wisconsin distribution, 23 June-17 August.
Resh, 1975. Kentucky distribution, 10 June-16 August.
Resh, *et al.,* 1975. Kentucky distribution (Salt River), June; August.
Roy and Harper, 1975. Quebec distribution, 11 July-29 August.
Etnier and Schuster, 1979. Tennessee distribution, 6 June-16 August.
Masteller and Flint, 1979. Pennsylvania distribution.
Neves, 1979. Massachusetts distribution, June to September.
Roy and Harper, 1979. Quebec distribution, 12 June-31 August.
Wimmer, 1979. Vermont distribution, 2 July-1 August.
Mingo and Gibbs, 1980. Maine distribution (Narraguagus River), 4-12 July.
Morse, *et al.,* 1980. South Carolina distribution, May-August.
Parker and Voshell, 1981. Virginia checklist.
Holzenthal, *et al.,* 1982. Louisiana distribution, 9 April-13 October
Unzicker, *et al.,* 1982. North and South Carolina distribution, May-mid-September.
Hamilton, *et al.,* 1983. Kansas distribution.
Harris, *et al.,* 1983. Alabama distribution (Tyro Creek watershed), May-July; September.
Huryn and Foote, 1983. Ohio distribution, June-September.
Waltz and McCafferty, 1983a. Indiana distribution, 10 July-5 September.
Haag, *et al.,* 1984. Kentucky distribution (Salt River).
Harris, *et al.,* 1984. Alabama distribution (Cahaba River system), May-October .
Lake, 1984. Delaware distribution, June-August.
Bowles and Mathis, 1989. Arkansas distribution, May-July; September.
Morse, *et al.,* 1989. North and South Carolina distribution (Lake Jocassee Catchment), May; July-August.
Floyd and Schuster, 1990. Kentucky distribution (Buck Creek system), June-July.
Glover and Tarter, 1990. West Virginia distribution, June-August.
Harris and Lago, 1990. Alabama distribution, April-October.
Tarter, 1990. West Virginia distribution, June-August.
Frazer, *et al.,* 1991. Alabama distribution (Little River drainage), May-September.
Harris, *et al.,* 1991. Alabama distribution, April-October.
Bowles and Mathis, 1992. Oklahoma distribution, May; August.
Masteller and Flint, 1992. Pennsylvania checklist.
Mathis and Bowles, 1992. Missouri distribution, May, August.
Floyd, *et al.,* 1993. South Carolina distribution (Upper Three Runs Creek), April-October.

DESCRIPTION. Larva: Length 6.0-7.0 mm. Head yellow, with dark muscle scars and small spots on frontoclypeus and parietal sclerites (Figure 6A); antennae short, reaching posterior edge of labrum; ventral apotome light yellow, saucer-shaped, with small, often indistinct, posterolateral triangular sclerites (Figure 6C). Pronotum with faint muscle scars laterally and posterolaterally (Figure 6B); remainder of morphology identical to that of *O. georgia.* **Case:** Length up to 8.0 mm, otherwise identical to that of *O. georgia* (Figure 5D).

MATERIAL EXAMINED. ALABAMA: Bibb Co., Cahaba R. @ Rt. 26, 7 L, 18 June 1992, MAF (CUAC); Little Cahaba R. @ Rt. 65, 2 L, 12 Mar. 1992, MAF and JBG (CUAC); Schultz Cr., @ Rt. 5, 3 L, 12 Mar. 1993, MAF (CUAC); same but 7 L, 18 June 1992; Schultz Cr. @ Rt. 219, 19 L, 12 Mar. 1992, MAF and JBG (CUAC); same but 28 L, 18 June 1992, MAF; Six Mile Cr. @ Rt. 25, 4 L, 12 Mar. 1992, MAF and JBG (CUAC); Escambia Co., Burnt Corn Cr. @ Rt. 41, 9 L, 15 Dec. 1991, MAF (CUAC); Lauderdale Co., stream @ Rt. 50 Br., 3 L, 2 Nov. 1990, MAF (CUAC); Lee Co., Little Uchee Cr. @ Rt. 178, Meadow's Mill, 10 L, 19 June 1991, MAF (CUAC); Tallapoosa Co., str. @ Rt. 258, 1 L, 4 Nov. 1990, MAF (CUAC); **ARKANSAS:** Clark Co., L'eau Frais Cr., @ Rt. 7, 1 L, 10 Mar. 1993, MAF (CUAC); Caddo R., Rt. 67, 5 L, 10 Mar. 1993, MAF (CUAC); Independence Co., White R., 2.3 mi. NE

Boise, 1 L, 31 July 1974, J. Richardson (ANSP); Montgomery Co., Little Missouri R., 9 L, 1980, J. F. (CUAC); Pike Co., Saline Cr. @ Rt. 26, 3 L, 10 Mar. 1993; Caddo R. @ Rt. 70, Glennwood, 1 L, MAF (CUAC); **FLORIDA:** Alachua Co., Santa Fe R., 1 L, 14 July 1952, W. Beck, Jr., (FAM); Clay Co., Black Cr., 1 L, 4 Jan. 1952, W. Beck, Jr. (FAM); unnamed str., St. John's Basin, 1 L, 25 June 1951, W. Beck, Jr. (FAM); Columbia Co., Santa Fe R. @ Rt. 441, 12 L, 17 Dec. 1991, MAF (CUAC); Gadsden Co., Telogia Cr. @ Rt. 650, 1 L, 1 July 1987, M. Pescador (FAM); Quincy Cr. @ Quincy, 1 L, 23 Sept. 1987, M. Pescador (FAM); Hamilton Co., Suwannee R., 5 L, 10 June 1953, W. Beck, Jr. (FAM); same but 1 L, 10 Mar. 1953; Jefferson Co., Burnt Mill Cr. @ Rt. 41, 85 L, 16 Dec. 1991, MAF (CUAC); Burnt Mill Cr. @ Rt. 59, 9 L, 31 July 1991, Hubbard, Pescador, and Silvanima (FAM); same but 2 L, 29 June 1991, Jones, Hubbard, and Silvanima; Jefferson/Taylor Co., Aucilla R. @ Rt. 98, 11 L, 19 June 1991, Hubbard and Silvanima (FAM); Lafayette Co., small str., 1 L, 7 July 1953, W. Beck, Jr. (FAM); Marion Co., Oklawaha R., 4 L, 17 Mar. 1952, W. Beck, Jr. (FAM); Orange Co., Little Econlockhatchee R. @ Barry Deese Rd., 3 L, 4 Dec. 1991, M. Vogel (CUAC); Big Wekiwa R., 1 L, 17 Sept. 1992, M. Vogel (CUAC); Big Wekiwa R. @ Wekiwa Marina, 3 L, 15 Mar. 1992, M. Vogel (CUAC); Big Econlockhatchee R. @ Altman, 7 L, 6 Aug. 1992, M. Vogel (CUAC); Orange/Osceola Co., Boggy Cr., 1 L, 13 Mar. 1992, M. Vogel (CUAC); Pasco Co., Withlacoochee R., 2 L, 11 May 1954, W. Beck, Jr. (FAM); Polk Co., Bear Br. str., 13 L, 9 Jan. 1951, W. Beck, Jr. (FAM); Taylor Co., Encofina R. @ Rt. 98, 3 L, 19 June 1991, Hubbard and Silvanima (FAM); Fenholloway R., 1 L, 12 Sept. 1951, W. Beck, Jr. (FAM); Wakulla Co., Wakulla R. @ Wakulla, 3 L, 13 Jan. 1959, M. Collins (INHS); **INDIANA:** Ripley Co., Laughery Cr. @ Versailles St. Pk., 1 L, 18 Mar. 1973, McCafferty, Provonsha, and Black (PU); **KENTUCKY:** Pulaski Co., Buck Cr. @ Rt. 39, 4 L, 9 Aug. 1991, MAF and N. Floyd (CUAC); Buck Cr. @ Rt. 1677, 3 L, 11 Aug. 1991, MAF and C. Floyd (CUAC); Spencer Co., Salt R. @ Rt. 33, 1 L, 10 Aug. 1989, J. Glover (CUAC); Salt R. @ Rt. 625, 2 L, 2 Aug. 1989, J. Glover (CUAC); Salt R. @ Rt. 623, 7 L, 27 May 1992, MAF and JBG (CUAC); **MAINE:** Kennebec Co., Kennebec R. @ Pishon Ferry, 2 L, 15 Aug. 1975, J. Richardson (ANSP); **MARYLAND:** Frederick Co., Potomac R., Monocacy, 4 L, 21 Aug. 1975, S. Roback (ANSP); same but 11 L, 22 Aug.; Potomac R., point of rocks, 1 mi. above Rt. 15, 2 L, 9 June 1961, S. Roback (ANSP); same but 4 L, 24-25 Aug. 1957; Montgomery Co., Potomac R., 4 mi. below mth. Monocacy, 1 L, 10 June 1961, S. Roback (ANSP); same but 2 L, 20-21 Aug. 1957; Potomac R. @ mth. Monocacy R., 3 L, 23 Aug. 1975, S. Roback; Potomac R. @ mth. Watts Br., 1 L, 24 May 1977, J. Richardson (ANSP); Potomac R., 2.5 mi. WSW Dickerson, 1 L, 6 June 1973, J. Richardson (ANSP); Potomac R. @ Mason Isl., 2 L, 25 Aug. 1975, S. Roback (ANSP); Potomoc R., 1.5 mi. below White's Ferry, l L, 26 Aug. 1975, J. Richardson (ANSP); same but 4 L, 22-23 Aug. 1957, S. Roback (ANSP); same but 2 L, 22 Aug. 1961, S. Richardson (ANSP); Worcester Co., Potomac R., Pocomoke City, Rt. 675, 1 L, 2 Feb. 1989, OSF (USNM); same but 6 July; **MICHIGAN:** Kalamazoo Co., August Cr. in Kellog Forest, 1 L, 12 Aug. 1969, GBW (ROM); Gull Cr. @ Crum Pk., 2 L, 12 Aug. 1969, GBW (ROM); Oakland Co., Huron R., Pontiac Lk. St. Pk., 3 L, 4 July, 1992, D. Bidlack (CUAC); **MINNESOTA:** Clearwater Co., Mississippi R., 5 mi. N Lk. Itasca P. O., 2 L, 28 July-19 Aug. 1970, GBW (ROM); Hubbard Co., Birch Cr. @ Rt. 4, N Lk. George, 2 L, 15 Aug. 1974, GBW (ROM); Lake Co., Kawishiwi R., Superior Nat. For., 1 L, 8 June 1936, R. Daggy (UM); LaSalle Cr., .5 mi. E Rt. 200, 1 L, 2 June 1981, P. Schefter and E. Fuller (ROM); Pine Co., Snake R., 2 L, 29 May 1977, G. Daussin (UM); **NEW JERSEY:** Ocean Co., Wrangle Brook @ Lakehurst, 1 L, 16 May 1962, O. Flint (USNM); **NORTH CAROLINA:** Alamance Co., Town Cr. @ Rt. 1200, 1 L, 5 May 1992, D. Lenat (NCDEM); Brunswick Co., Waccamaw R. @ Freeland, 3 L, 7 June 1987, D. Lenat (NCDEM); Columbus Co., Waccamaw R. , 9 L, 19 June 1991, D. Lenat (NCDEM); Durham Co., Little R. nr. orange factory, 4 L, 21 July 1987, D. Lenat (NCDEM); Edgecombe Co., Fishing Cr., nr. Enfield, 5 L, 24 July 1985, D. Lenat (NCDEM); Harnett Co., Cape Fear R. @ Lillington, 15 L, 12 Sept. 1990, D. Lenat (NCDEM); same but 7 L, 8 July 1986, D. Lenat (NCDEM); Johnston Co., Marks Cr. @ Rt. 1714, 5 L, 15 July 1991, D. Lenat (NCDEM); Jones Co., Mill Run, 4 L, July 1991, D. Lenat (NCDEM); Madison Co., French Broad R., Marshall, 15 L, July 1992, D. Lenat (NCDEM); Nash Co., Tar R. @ Rocky Mount, 6 L, 12 July 1990, D. Lenat (NCDEM); Onslow Co., New R. @ Gum Br., 13 L, 24 June 1986, D. Lenat (NCDEM); Pitt Co., Contantnea Cr., 1 L, 22 July 1991, D. Lenat (NCDEM); Robeson Co., Lumber R., 8 L, 11 Sept. 1991, D. Lenat (NCDEM); Swain Co., Deep Cr. @ Deep Cr. Missionary Bapt. Ch., 2 mi. N Bryson City, 4 L, 14 July 1982, Hamilton, Rothschild, and Morse (CUAC); Wake Co., Neuse R. @ Rt. 64, 5 L, 10 July 1991, D. Lenat (NCDEM); Wayne Co., Little R., 2 L, 24 July 1991, D. Lenat (NCDEM); **OHIO:** Ashland Co., Clear Fk. R. @ Mohican St. Pk., 8 L, 2 Aug. 1968, TY and L. Kohalmi (ROM); **OKLAHOMA:** Pushmataha Co., Kiamichi R. @ Tuskahoma, 2 L, 11 Mar. 1993, MAF (CUAC); **ONTARIO:** Algonquin Prov. Pk., Source Lk., 1 L, 2 July 1954, GBW (ROM); Costello Cr. @ Opeongo Lk. Rd., .5 mi. N Rt. 60, Algonquin Prov. Pk., 1 L, 11 May 1972, ROMFP (ROM); Madawaska R. @ Rt. 60, Algonquin Prov. Pk., 7 L, 27-28 Sept. 1972, ROMFP (ROM); same but 46 L, 31 May 1972; Papineau Cr., Rt. 127 NW Maynooth, 2 L, 9 May 1972, ROMFP (ROM); Algoma D., Barrett R., Rt. 17, 79 km S Wawa, 1 L, 30 June 1981, C. Parker (ROM); Durham Co., N Ganaraska R., crossing 5th line W Perrytown, 1 L, 19 June 1973, GBW and TY (ROM); Haliburton D., Boshkung R., Sherborne Portage, 3 L, 1 Sept. 1977, P. Schefter (ROM); str. below beaver dam @ L. Kawagama, Dorset, 3 L, GBW, Sutton, and Odum (ROM); Hastings

Co., Moira R., Latta, 2 L, 20 June 1972, ROMFP (ROM); Muskoka D., Bala, culvert where Hart L. flows into Muskoka R., 1 L, 27 July 1972, B. Smith (ROM); Nipissing D., Algonquin Prov. Pk., Madawaska R., 1 L, 29 June 1973, I. and S. Smith (ROM); str. beside Rt. 11, 3 mi. S Trout Cr., 2 L, 1 June 1972, ROMFP (ROM); same but 14 May 1972; Parry Sound D., Gurd Lk., Grundy Lk. Prov. Pk., Rt. 69 S Sudbury, 5 L, 1 June 1971, ROMFP (ROM); Peterborough Co., Indian R. @ Warsaw Caves Cons. Area, 4 L, 6 Oct. 1969, I. Smith (ROM); Thunder Bay D., Creelman Cr., crossing Rt. 11, 13.6 mi. SW Geraldton, 1 L, 22-23 June 1971, ROMFP (ROM); Timiskaming D., Cripple Cr., Rt. 101, 29 mi. W Timmins, 2 L, 19 May 1972, ROMFP (ROM); Kelley Cr., Rt. 101, 28.8 mi. W Timmins, 1 L, ROMFP (ROM); **PENNSYLVANIA:** Delaware Co., Ridley Cr. nr. Westchester Pike, 1 L, 12 Aug. 1952, S. Roback (ANSP); same but 8 June 1959; Wyoming Co., Susquehanna R., 2 mi. S. Meshoppen, 1 L, 16 Aug. 1977, J. Richardson (ANSP); **SOUTH CAROLINA:** Aiken Co., Boggy Gut Cr. @ Rd. 781-4, Sav. R. Site, 1 L, 14 Dec. 1976, Herlong and Prichard (CUAC); Mill Cr. @ Rd. E-2, Sav. R. Site, 2 L, 8 July 1977, Herlong and Prichard (CUAC); same but 17 May; same but 1 L, 24 Mar.; same but 1 June; same but 8 Aug.; same but 14 Dec. 1976; Tinker Cr. @ Rd. 8-1, Sav. R. Site, 1 L, 8 July 1977, Herlong and Prichard (CUAC); same but 17 May 1977; Upper Three Runs Cr. @ Rd. C, 2 L, 22 July 1971, J. Richardson (ANSP); same but 1 L, 8 Aug. 1977, Herlong and Prichard (CUAC); Anderson Co., Little Garvin Cr. @ Rt. 57, 1 L, 13 Oct. 1978, T. White (CUAC); Watermelon Cr., 1 L, 22 Sept. 1979, J. Weaver (CUAC); Berkeley Co., Four Holes Swamp, Mims Lake, 2 L, Oct. 1979, S. Hamilton (CUAC); Chesterfield Co., Fork R. @ Rt. 624, Brewer Gold Mine, 1 L, 29 Mar. 1991, Heyn, Shealy, Bouknight (CUAC); Darlington Co., Black Cr., just above Darlington, 1 L, 11 Sept. 1972 (ANSP); Edgefield Co., unnamed trib. Beaverdam Cr., 10 mi. W Edgefield, 16 L, 23 July 1993, MAF (CUAC); Kershaw Co., Wateree R. nr. Camden, 2 L, 31 Aug. 1958, S. Roback (ANSP); Laurens Co., Enoree R. @ Rt. 56, 20 L, 22 July 1991, MAF and K. Moulton (CUAC); Oconee Co., Chattooga R. @ Rt. 28, 1 L, 10 May 1991, F. Winborne (CUAC); same but 27 L, MAF; same but 6 L, L. Eaton; same but 1 L, 11 May, C. Wells; same but 5 L, 14 June 1991, MAF; Coneross Cr. @ Rt. 59, 1 L, May 1981, S. Hamilton (CUAC); Little R. @ Rt. 24, 3 L, 17 Apr. 1991, MAF (CUAC); same but 3 L, 29 Apr. 1992; Orangeburg Co., N Fk. Edisto R., downstr. Glover St., Orangeburg, 5 L, 21 June 1991, M. Heyn (CUAC); Pickens Co., Rocky Bottom Cr. @ Rt. 100, 2 L, 22 Sept. 1991, MAF (CUAC); same but 2 L, 21 July 1991, MAF and J. Moulton; same but 1 L, 20 July; Six Mile Cr. @ Rt. 219, 10 L, 20 July 1991, MAF and J. Moulton (CUAC); Spartanburg Co., Middle Tyger R. @ Rt. 63, 1 L, 22 Aug. 1991, M. Heyn (CUAC); N Tyger R. @ ABCO Chem., 1 L, 3 May 1990, M. Heyn (CUAC); Pacolet R. @ Rt. 31, 1 L, 27 July 1991, MAF and J. Moulton (CUAC); Pacolet R. @ Rt. 737, 1 L, 27 July 1991, MAF and J. Moulton (CUAC); **TEXAS:** ?, Guadalupe R. nr. Victoria, 1 L, 19 Aug. 1952, T. A. (ANSP); **VIRGINIA:** Page Co., S Fk. Shenandoah R., 10 L, 19 July 1992, J. Glover (CUAC); Scott Co., Clinch R., 2 L, 1 Aug. 1954, S. Roback (ANSP); Washington Co., Holston R. below Rt. 91 Br., 1 L, 26 June 1974, S. Roback (ANSP); ?, James R. nr. Richmond, 1 L, 26 July 1951, J. Latlin (ANSP); **WEST VIRGINIA:** Wayne Co., Big Sandy R., mi. 16.8 nr. Prichard, 2 L, 11 Aug. 1987, R. Kirchner (USNM); **WISCONSIN:** Florence Co., Pine R., end 2nd Rd. S jct. Co. Rt. N & D, 2 L, 4 June 1981, P. Schefter and E. Fuller (ROM); Washburn Co., Namekagon R. @ Spooner, 9 L, 5-6 June 1936, Frison and Ross (INHS).

ASSOCIATION, BIOLOGICAL NOTES, AND DISTRIBUTION. The association of larvae and adults for *O. persimilis* was based on rearings from four South Carolina streams, Chattooga River (Oconee County), Little River (Oconee County), Pacolet River (Spartanburg County), and Rocky Bottom Creek (Pickens County) (List of Reared Specimens, p. 39). Larvae have been found in lotic environments ranging in size from small streams (<1 m wide) to large rivers (> 50 m wide). Larvae are collected most frequently from root mats and woody debris but also can be found at the base of aquatic macrophytes such as water willow, *Justicia americana* (L.) Vahl. Like *O. cinerascens* and *O. georgia*, this species is able to swim due to the oar-like function of the swimming hairs on the hind tibiae. As shown by the extensive material examined, larvae of *O. persimilis* are more commonly collected than those of anyother *Oecetis* species. This species is widely distributed in eastern North America (Map 3). Its range includes Alabama, Arkansas, Delaware, Florida, Georgia, Illinois, Indiana, Kansas, Kentucky, Louisiana, Massachusetts, Maine, Michigan, Minnesota, Mississippi, Missouri, New Hampshire, New Jersey, New York, North Carolina, Ohio, Oklahoma, Ontario, Pennsylvania, Quebec, South Carolina, Tennessee, Texas, Vermont, Virginia, West Virginia, and Wisconsin.

DIAGNOSIS. *Oecetis persimilis* is morphologically most similar to *O. georgia* but differs from it in the presence of dark muscle scars on the head. The muscle scars are very faint or absent on *O. georgia*.

OECETIS (QUARIA) MILNE 1934

Oecetis (Quaria) scala Group

Oecetis eddlestoni Ross (Figure 8, Map 4)

Oecetis eddlestoni Ross. 1938. Bull Illinois Nat. Hist. Surv. 21:160 (Type locality: Pennsylvania, Bradford Co., Susquehanna R. @ Sayre, 29 July 1937, H. Ross).

Unzicker, *et al.,* 1970. Arkansas distribution, May-August.
Hamilton, *et al.,* 1983. Kansas distribution.
Huryn and Foote, 1983. Ohio distribution, 1 June-26 August.
Bilger, 1986. New England checklist (Vermont).
Bowles and Mathis, 1989. Arkansas distribution, May-August.
Bowles and Mathis, 1992. Oklahoma distribution, May-August.
Masteller and Flint, 1992. Pennsylvania checklist.
Moulton and Stewart, 1993. Discussion of *O. scala* Group.

DESCRIPTION. Larva: Length 9.0 mm. Head yellow, with numerous muscle scars on frontoclypeus and parietal sclerites (Figure 8A); postgenal parietal sclerites divided by dark and pale areas (Figure 7C), subocular line indistinct; antennae short, reaching posterior margin of labrum; ventral apotome 4 times as wide as long, with small, posterolateral, triangular sclerites (Figure 8C). Pronotum yellow, with muscle scars along posterior margin of darkened area (Figure 8B); additional muscle scars at posterolateral corners; row of long setae along anterior and lateral margins; transverse row of 5-6 setae posterior to anterior row on each side, setae each with circular, dark area at base. Mesonotum light yellow, with 1 seta at each anteromesal corner; 8-10 setae at each anterolateral corner, 2 setae on middle of each side. Metanotum membranous, each side with 1 Sa2 seta and 1 Sa3 seta. Mesosternum with 1 pair of setae. Foretrochantins each with 1 dorsal seta. Claws of each leg with 1 basal seta; midlegs and hindlegs with tarsi longer than claws. Each side of abdomen with 1 dorsal and 1 ventral gill on segments II to VII; lateral sclerite of segment I light yellow, each with 2 ventral setae; abdominal tergite IX with 6 large setae and 4 small setae. **Case:** Length 9.0 mm. Tubular in shape, composed of sand grains (Figure 8D); slightly dorsoventrally compressed, with few large sand grains on lateral margins; slightly tapered with some curvature.

MATERIAL EXAMINED. ILLINOIS: LaSalle Co., Indian Cr. @ Serena, 10 L, 12 May 1938, Ross and Burks (INHS); same but 1 L, 27 May 1938.

ASSOCIATION, BIOLOGICAL NOTES, AND DISTRIBUTION. Positive adult and larval associations for this species were accomplished by Herbert Ross of the Illinois Natural History Survey in 1938. He reared at least four individuals in cages placed in Indian Creek, Spring Grove, Illinois. Since that time, no additional larval specimens of this species have been collected. Consequently, little is known of its biology. Because Ross' specimens have deteriorated considerably over the last 56 years, color patterns of sclerites on the head and pronotum were difficult to discern. Therefore, the provided illustrations should be considered tentative. Adults have been collected infrequently and in small numbers from May to August. This species has a widely scattered distribution and is known only from seven states: Arkansas, Illinois, Kansas, Ohio, Oklahoma, Pennsylvania, and Vermont (Map 4).

DIAGNOSIS. *Oecetis eddlestoni* can be distinguished from the other associated species of the *scala* Group by its one pair of metasternal setae and its two ventral setae on each lateral sclerite of abdominal segment I.

Oecetis morsei Bueno-Soria (Figure 7, Map 5)

Oecetis morsei Bueno-Soria. 1981. Folia Entomol. Mex. 49:116-119 (Type locality: South Carolina, Aiken Co., Upper Three Runs Cr., 28 June 1977, Herlong and Prichard).

Morse, *et al.,* 1980. South Carolina distribution (Upper Three Runs Creek), May-June; August-September.
Harris, *et al.,* 1982. Florida distribution (Eglin Air Force Base), 20 May.

Gordon, 1984. Florida distribution.
Harris, *et al.,* 1984. Alabama distribution (Cahaba River system), June-October.
Harris, 1990. Alabama rare and endangered invertebrates.
Harris and Lago, 1990. Alabama distribution, June; August; October.
Harris, *et al.,* 1991. Alabama distribution, June; August; October.
Floyd, *et al.,* 1993. South Carolina distribution (Upper Three Runs Creek), May-October.
Moulton and Stewart, 1993. Discussion of *O. scala* Group.

DESCRIPTION. Larva: Length 5.0 mm. Head yellow-brown, with many dark muscle scars and spots on frontoclypeus and parietal sclerites (Figure 7A); irregular, longitudinal, dark stripes present along frontoclypeal sutures mesad to eyes, connected posteriorly by transverse stripe, together forming a p-shaped (pi-shaped) pattern; postgenal parietal sclerites divided into distinct, yellow-brown and pale areas (Figure 7C), coronal suture indistinct; antennae reaching just beyond posterior edge of labrum; ventral apotome trapezoidal, with large, posterolateral, subtriangular sclerites (Figure 7D). Pronotum yellow-brown on anterior two-thirds, with muscle scars present laterally and mesally along posterior edge (Figure 7B); row of long setae along anterior and lateral margins; transverse row of 6 setae posterior to anterior row on each side, bases of setae each with oval, dark, spot; 5 -6 muscle scars present at each posterolateral corner. Mesonotum light brown, with 1 seta near each anteromesal corner; 8-10 setae present on each anterolateral corner, 2 setae on middle of sclerites. Metanotum with 1 Sa2 seta and 1 Sa3 seta on each side. Metasternum with transverse row of 12 setae. Foretrochantins each with 1 dorsal seta. Tarsi of legs longer than respective claws; claws each with 1 basal seta. Each side of abdomen with 1 dorsal and 1 ventral gill on segments II to VII, 1 pair of pleural gills on segment II; lateral sclerites of segment I brown and elongate, each with l seta on ventral margin (Figure 7E); tergite of segment IX with 6 large setae, 4 small setae. Each anal proleg with brown basal sclerite. **Case:** Length 5.0 mm. Tubular in shape, composed of sand grains (Figure 7F); tapered with some curvature, especially early instars.

MATERIAL EXAMINED. SOUTH CAROLINA: Aiken Co., Upper Three Runs Cr. @ Rd. 8-1, 1 L, 8 Aug. 1977, Herlong and Prichard (CUAC); same but 4 May 1977; same but 22 July; same but 1 June; same but 14 June; same but 17 May; same but 13 June; same but 9 L, 28 June; same but 6 L, 8 July; same but 1 L, 27 June 1979, Kelley and McEwan; same but 8 L, 10 July 1979; same but 1 L, 4 Apr. 1980, McEwan and Powell; same but 3 L, 11 May 1979; same but 5 L, 11 June; same but 3 L, 23 May; same but 1 L, 31 Mar.; same but 18 June 1991, MAF; Upper Three Runs Cr. @ Rt. 278, 4 L, 27 Apr. 1992, MAF (CUAC); Mill Cr. @ Rd. E-2, Sav. R. Site, 1 L, 8 June 1977, Herlong and Prichard (CUAC).

ASSOCIATION, BIOLOGICAL NOTES, AND DISTRIBUTION. Association of larvae and adults was based on rearings from Upper Three Runs Creek, a fourth order, swift, sand-botttomed, blackwater stream in Aiken County, South Carolina (List of Reared Specimens, p. 39). This species occurs strictly in lotic environments, where it lives on aquatic macrophytes or submerged roots of riparian vegetation (root mats). Unlike other root-dwelling species (*e.g., O. georgia*), it does not build a plant case and lacks the ability to swim. During pupation, the case is secured with silk to pieces of roots or macrophytes. In South Carolina, the adults emerge from May to October. *Oecetis morsei* is collected infrequently and known from only three southeastern states: Alabama, Florida, and South Carolina. It has a very localized distribution and has been collected abundantly only in the Upper Three Runs Creek drainage of Barnwell and Aiken Counties, South Carolina (Map 5). Harris (1990) considered *O. morsei* to be rare or threatened in Alabama.

DIAGNOSIS. Presently, the larva of *O. morsei* cannot be distinguished from the larva of *O. sphyra*. It differs from the other associated member of the *scala* Group, *O. eddlestoni*, by the presence of one seta on each lateral sclerite of abdominal segment I and the transverse row of 12 setae on the metasternum.

Oecetis ouachita Moulton (Map 6)

Oecetis ouachita Moulton and Stewart. 1993. Ann. Entomol. Soc. Am. 86:222 -226 (Type locality: Arkansas, Ouachita Co., Ouachita River @ Tate's Bluff Access & Little Missouri River confluence, 8 km NE Childester and 8.7 km SW Sparkman, 4 July 1987, E. J. Bacon).

Bowles and Mathis, 1989. Arkansas distribution (*O. scala*, 1 male), July.

BIOLOGICAL NOTES AND DISTRIBUTION. Almost nothing is known of this species' biology. Nonetheless, the larvae are thought to be strictly lotic, because adults have been collected only near large rivers. I attempted in March 1993 to collect larvae of this species but was unsuccessful, primarily due to high water levels. However, one empty pupal case was found among roots at the type locality. The case was identical to that of *O. morsei* and *O. sphyra*. Adults of this species have been collected in very large numbers (some samples with > 100 individuals) from 13 June to 18 July (Moulton and Stewart,1993). This species is restricted to Arkansas and Oklahoma (Map 6). As reported by Moulton and Stewart (1993), "*O. ouachita* is known only from large river systems in the Ouachita Mountains of Arkansas and Oklahoma, including the bordering Gulf Coastal Plain region in Arkansas."

Oecetis ozarkiensis Moulton (Map 7)

Oecetis ozarkiensis Moulton and Stewart. 1993. Ann. Entomol. Soc. Am. 86:222-226 (Type locality: Arkansas, Stone Co., Gunner Pool Rec. Area, North Sylamore Cr., 22 July 1987, C. E. Carlton).

Mathis and Bowles, 1992. Missouri distribution (*Oecetis* sp. nr. *eddlestoni*, 1 male), August.

BIOLOGICAL NOTES AND DISTRIBUTION. I attempted in March 1993 to locate the larva of this species but was unsuccessful. Little is known concerning its biology. Moulton and Stewart (1993) described the type locality as "a highly enriched lentic habitat formed by the construction of a small dam on a first order tributary of North Sylamore Creek," with a riparian canopy dominated by an oak-hickory-pine forest. A total of 17 adult males have been collected from June to August (Moulton and Stewart, 1993). This species is restricted to the type locality and another location in south-central Missouri (Christian County, Finley Creek) (Moulton and Stewart, 1993) (Map 7). Efforts by S. R. Moulton, II, to locate the specimen reported by Mathis and Bowles (1992) from Reynolds County, Missouri, were unsuccessful.

Oecetis scala Milne (Map 8)

Oecetis scala Milne. 1934. Stud. N. Am. Trichoptera 1:17 (Type locality: Maryland, Del. Water Gap).

Roy and Harper, 1975. Quebec distribution, 27 July.
Roy and Harper, 1979. Quebec distribution, 27 July.
Unzicker, *et al.,* 1982. North and South Carolina distribution, early May.
Harris, *et al.,* 1980. Alabama distribution, June-October.
Harris and Lago, 1990. Alabama distribution, May-October.
Frazer, *et al.,* 1991. Alabama distribution, June-July; September.
Harris, *et al.,* 1991. Alabama distribution, May-October.

BIOLOGICAL NOTES AND DISTRIBUTION. Little is known concerning the biology of this species. Adults have been collected from May-October and found most abundantly in two Alabama river systems, Cahaba River of central Alabama (Harris, *et al.,* 1991) and Little River, Cherokee County, northeastern Alabama (Frazer, *et al.,* 1991). Attempts in March 1992 and 1993 to find the larva of *O. scala* from localities within the Cahaba River system were unsuccessful. Early instar larvae of the *O. scala* Group were found that were morphologically identical to *O. morsei* and *O. sphyra*. However, I am unsure if these larvae were *O. scala*, because they were too immature for successful rearing. Based on the similarity of some adults and larvae within the *O. scala* Group (*i.e., O. morsei*, *O. ouachita*, *O. scala*, and *O. sphyra*), the larva of *O. scala* may be morphologically inseparable from other members of the group. *Oecetis scala* has an eastern distribution that generally follows the Appalachian Mountains (Harris, *et al.,* 1991) (Map 8). It is known from Alabama, Maryland, New Hampshire, New Jersey, North Carolina, Pennsylvania, and Quebec.

Oecetis sphyra Ross (Map 9)

Oecetis sphyra Ross. 1941. Trans. Amer. Entomol. Soc. 67:99 (Type locality: Georgia, Upson County, Thomaston, 8 June 1939, P. Fattig).

Etnier and Schuster, 1979. Tennessee distribution, 8 June-1 July.
Parker and Voshell, 1981. Virginia checklist.
Harris, *et al.,* 1982. Florida distribution (Eglin Air Force Base), 8 June, 19 September.
Holzenthal, *et al.,* 1982. Louisiana and Mississippi distribution, 23 May-18 October.
Unzicker, *et al.,* 1982. North and South Carolina distribution, June-August.
Harris, *et al.,* 1983. Alabama distribution (Tyro Creek watershed), June.
Harris, *et al.,* 1984. Alabama distribution (Cahaba River system), June-August.
Morse, *et al.,* 1989. South Carolina distribution (Lake Jocassee Catchment), June.
Harris and Lago, 1990. Alabama distribution, April-September.
Harris, *et al.,* 1991. Alabama distribution, April-September.
Moulton and Stewart, 1993. Discussion of *O. scala* Group.

DESCRIPTION. Larva: Identical to that of *O. morsei.* **Case:** Identical to that of *O. morsei.*

MATERIAL EXAMINED. NORTH CAROLINA: Gaston Co., S Fk. Catawba R. @ McAdenville, 5 L, 18 Aug. 1992, D. Lenat (NCDEM); Nash Co., Swift Cr., 1 L, 10 May 1991, D. Lenat (NCDEM); Wake Co., Neuse R. @ Rt. 64, 2 L, 10 July 1991, D. Lenat (NCDEM); **SOUTH CAROLINA:** Laurens Co., Enoree R. @ Rt. 56, 1 L, 22 July, 1991, MAF and J. Moulton (CUAC); Oconee Co., Coneross Cr. @ Rt. 59, 2 L, 18 May 1981, S. Hamilton (CUAC); Spartanburg Co., Middle Tyger R., downstr. Lyman Waste Water Tr. Plt., 1 L, 22 Aug. 1991, M. Heyn (CUAC); Pacolet R. @ Rt. 31, 25 L, 27 July 1991, MAF and J. Moulton (CUAC).

ASSOCIATION, BIOLOGICAL NOTES, AND DISTRIBUTION. The larval identity of *O. sphyra* was discovered through rearings from the Pacolet River, Spartanburg County, South Carolina (List of Reared Specimens, p. 39). At this collection site the Pacolet River is swift and clear, with a substrate of fine gravel and sand and a channel width of approximately 15 m. *Oecetis sphyra* is another strictly lotic species whose larvae inhabit the submerged roots of riparian vegetation and woody debris. In Alabama, adults were collected most abundantly along sand-bottomed streams of the Coastal Plain (Harris, *et al.,* 1991). Within the Cahaba River system in central Alabama, adults of *O. sphyra* were captured together with *O. morsei* and *O. scala* at three localities (Harris, *et al.,* 1984). In North and South Carolina, adults have been collected primarily from Piedmont rivers and streams. This species often occupies the same microhabitat (root mats) as *O. cinerascens*, *O. georgia*, and *O. persimilis* but does not make a plant case and, because it lacks swimming hairs, does not have the ability to swim. *Oecetis sphyra* is primarily a southeastern species but is known from as far north as Culpepper County, Virginia (Map 9). Its distribution includes the following states: Alabama, Florida, Georgia, Louisiana, Mississippi, North Carolina, South Carolina, and Virginia.

DIAGNOSIS. Presently, *O. sphyra* cannot be distinguished from *O. morsei.* It differs from the other associated member of the *scala* Group, *O. eddlestoni*, by the presence of one seta on each lateral sclerite of abdominal segment I and a transverse row of 12 setae on the metasternum.

Oecetis (*Quaria*) *arizonica* Group

Oecetis arizonica **Denning (Figure 9, Map 10)**

Oecetis arizonica Denning. 1951. J. Kansas Entomol. Soc. 24:159 (Type locality: Gordon Canyon, Cochise Co., Arizona, 3 July 1950, C. P. Alexander).

Herrmann, *et al.,* 1986. Colorado distribution, 28 July.

DESCRIPTION. Larva: Length 6.7-7.5 mm. Head longer than wide, light brown with contrasting, dark musclescars on frontoclypeus and postgenal parietals (Figure 9A); antennae long, reaching anteriorly to anterior edge of labrum; ventral apotome dark brown, trapezoidal, with smaller, posterolateral, subtriangular sclerites (Figure 9C). Pronotum dark brown, with contrasting, pale color pattern posteriorly; muscles scars present posterolaterally (Figure 9B); row of many long setae along anterior and lateral margins; transverse row of 4-5 setae posterior to anterior row on each side. Mesonotum light brown, each half with as many as 8 setae along anterior and anterolateral margins; 2 setae on middle of each side. Metanotum light brown, each side with 1-3 sa2 setae, 4-7 sa3 setae. Claw of each foreleg and midleg with short basal seta. Each side of abdomen with 1 dorsal and 1 ventral gill on segments

II to VII, pair of pleural gills on segment II; each sclerite of lateral hump light yellow, with 1 ventral seta; tergum of segment IX with 5 to 6 setae along posterior margin; anal claws each with one accessory hook. **Case:** Length 7.5-8.0 mm. Plant and detrital material combined with sand grains (Figure 9D); slightly tapered in younger instars.

MATERIAL EXAMINED. ARIZONA: Santa Rita Mts., Madera Canyon, 4 L, 12-13 June 1968, Menke and Flint (USNM); Cochise Co., Chiricahua Mts., Rucker Canyon, 1 L, 8 June 1968, Menke and Flint (USNM); Cochise Co., spring-fed pond, 5 mi. W Portal, 1 L, 9 June 1961, Kingsolver (INHS); Coconino Co., Cave Cr. @ Herb Martyr Cpgrd., 7 mi. W Portal, 5 L, 23 June 1966, GBW (ROM); Coconino Co., W Frk. Oak Cr., 3 L, 5 Dec. 1986, GBW (ROM).

ASSOCIATION, BIOLOGICAL NOTES, AND DISTRIBUTION. At present, the larvae and adults of this species have not been associated by rearing. However, I have examined numerous larval specimens from Arizona that are morphologically very distinct from other North American species and that occur in the same areas where *O. arizonica* adults have been collected. Consequently, I believe this species is probably the larva of *O. arizonica*. The larva of this species has been found only in streams and rivers, many of which occur in mountainous regions. Adult collections have been made from June to July at altitudes over 2,000 m. The distribution of *O. arizonica* is restricted to two states, Arizona and Colorado (Map 10), but future collecting in adjacent states (*i.e.,* California, New Mexico) probably will extend its range. It also has been reported from Mexico (Flint, 1967).

DIAGNOSIS. This species can be differentiated from other North American species by the anterior, dark brown area of the pronotum.

OECETIS (*OECETIS*) MACLACHLAN 1877

Oecetis (*Oecetis*) *furva* Group

Oecetis ditissa Ross (Map 11)

Oecetis ditissa Ross. 1966. Ill. Acad. Sci. Trans. 59:13 (Type locality: Illinois, Jackson County, Grand Tower @ lights, 30 May 1935, H. Ross and C. Mohr).

Harris, 1971. Illinois distribution (Kaskaskia River), July-October.
Resh, 1975. Kentucky distribution, 2 June-19 September.
Resh, *et al.,* 1975. Kentucky distribution (Salt River), June-July; September.
Etnier and Schuster, 1979. Tennessee distribution.
Masteller and Flint, 1979. Pennsylvania distribution, 10 July.
Neves, 1979. Massachusetts distribution, August.
Morse, *et al.,* 1980. South Carolina distribution (Upper Three Runs Creek), September.
Parker and Voshell, 1981. Virginia checklist.
Holzenthal, *et al.,* 1982. Louisiana and Mississippi distribution, 15 April-17 October.
Unzicker, *et al.,* 1982. North and South Carolina distribution, mid-April-early September.
Hamilton, *et al.,* 1983. Kansas distribution.
Harris, *et al.,* 1983. Alabama distribution (Tyro Creek watershed), May-September.
Huryn and Foote, 1983. Ohio distribution, 31 July-14 September.
Waltz and McCafferty, 1983a. Indiana distribution, 30 May-3 October.
Gordon, 1984. Florida distribution.
Harris, *et al.,* 1984. Alabama distribution (Cahaba River system), May-October.
Usis and MacLean, 1986. Ohio distribution (Stillfork Swamp), 11 June-22 September.
Garono and MacLean, 1988. Ohio distribution (wetlands).
Bowles and Mathis, 1989. Arkansas distribution, May-July; September.
Floyd and Schuster, 1990. Kentucky distribution (Buck Creek system), May; July-September.
Harris and Lago, 1990. Alabama distribution, March-October.
Frazer, *et al.,* 1991. Alabama distribution (Little River drainage), May-October.
Harris, *et al.,* 1991. Alabama distribution, March-October.
Masteller and Flint, 1992. Pennsylvania checklist.

Bowles and Mathis, 1992. Oklahoma distribution, April-July; September.
Mathis and Bowles, 1992. Missouri distribution (Ozark Mountains), May-August.
Floyd, *et al.,* 1993. South Carolina distribution (Upper Three Runs Creek), April; August-October.
Moulton, *et al.,* 1993. Texas distribution (Brazos River drainage), April; June; October.

BIOLOGICAL NOTES AND DISTRIBUTION. The larval habitat of *O. ditissa* could consist of a variety of habitats, because adults have been collected near both lentic and lotic habitats. Due to the morphological similarity of their adults, larvae of *O. ditissa* and *O. nocturna* are possibly very similar in appearance. Larvae of the two species probably occur in the same habitats, because adults of both species frequently are collected together in the same light trap (Ross, 1966). When collected together, however, *O. ditissa* is often much less abundant (Haag, *et al.,* 1984, Frazer, *et al.,* 1991, Harris, *et al.,* 1991). This species is known from the eastern and midcentral United States (Map 11). Its range includes Alabama, Arkansas, Florida, Georgia*, Illinois, Indiana, Kansas, Kentucky, Louisiana, Massachusetts, Mississippi, Missouri, North Carolina, Ohio, Oklahoma, Pennsylvania, South Carolina, Tennessee, Texas, and Virginia.

Oecetis inconspicua (Walker)

Leptocerus inconspicua Walker. 1852. Cat. Br. Mus. Neur. 1:71-72 (Type locality: Georgia).

BIOLOGICAL NOTES AND DISTRIBUTION. Based mainly upon adult characteristics, *O. inconspicua* has long been suspected by many researchers to represent a species complex. The adults of *O. inconspicua* seem to occur everywhere, usually in large numbers. They have been collected throughout North America (primarily with ultraviolet light traps) from a diverse array of aquatic habitats (*e.g.,* bogs, artificial ponds, marshes, natural lakes, reservoirs, large rivers, and small streams). Most importantly, the adults have shown a tremendous amount of morphological variation (*e.g.,* overall size and genitalic structure).

The larval description by Ross (1944, Figure 834) also alluded to the possibility of a species complex by mentioning the presence of multiple larval forms. He stated that the larval head varied "from an almost immaculate condition to a distinct dark brown pattern." In further support of a species complex, I have made larval and adult associations for seven North American species (*Oecetis* sp. A-G, described below), that are morphologically distinct as larvae but represent "typical" *O. inconspicua* as adults. In addition, I have examined many other distinct, but morphologically similar, preserved specimens that also could be included in this complex. Both the actual number of species occurring in the complex and the larval identity of *O. inconspicua* remain to be discovered. A more thorough understanding of this complex will precede any attempt at a complete synopsis of the taxonomy, biology, and distribution of *Oecetis* species in North America.

Oecetis nocturna Ross (Figure 10, Map 12)

Oecetis nocturna Ross. 1966. Ill. Acad. Sci. Trans. 59: 11-13 (Type locality: Missouri, Taney Co., Hollister, 16 May 1938, V. Kite) .

Unzicker, *et al.,* 1970. Arkansas distribution, June.
Harris, 1971. Illinois (Kaskaskia River), July-October.
Resh, 1975. Kentucky distribution, 25 May-7 October.
Resh, *et al.,* 1975. Kentucky distribution (Salt River), May-October.
Roy and Harper, 1975. Quebec distribution, 25 August.
Roy and Harper, 1979. Quebec distribution, 29 June-25 August.
Resh, *et al.,* 1978. Oklahoma and Texas distribution.
Etnier and Schuster, 1979. Tennessee distribution.
Wimmer, 1979. Vermont distribution, 11 August.
Parker and Voshell, 1981. Virginia checklist.
Holzenthal, *et al.,* 1982. Louisiana and Mississippi distribution, 8 May-17 October.
Unzicker, *et al.,* 1982. North and South Carolina distribution, late April-early November.
Hamilton, *et al.,* 1983. Kansas distribution.
Harris, *et al.,* 1983. Alabama distribution (Tyro Creek watershed), May-September.

Huryn and Foote, 1983. Ohio distribution, 22 May-14 September.
Waltz and McCafferty, 1983a. Indiana distribution, 22 May-10 September.
Haag, *et al.,* 1984. Kentucky distribution (Salt River).
Harris, *et al.*, 1984. Alabama distribution (Cahaba River system), May-October.
Usis and MacLean, 1986. Ohio distribution (Stillfork Swamp).
Garono and MacLean, 1988. Ohio distribution (wetlands).
Bowles and Mathis, 1989. Arkansas distribution, June-September.
Morse, *et al.,* 1989. South Carolina distribution (Lake Jocassee Catchment), July.
Floyd and Schuster, 1990. Kentucky distribution (Buck Creek system), May; July-August.
Glover and Tarter, 1990. West Virginia distribution, June-July; September.
Harris and Lago, 1990. Alabama distribution, April-October.
Tarter, 1990. West Virginia distribution, June-July; September.
Frazer, *et al.,* 1991. Alabama distribution (Little River drainage), May-October.
Harris, *et al.,* 1991. Alabama distribution, April-October.
Bowles and Mathis, 1992. Oklahoma distribution, May-July.
Masteller and Flint, 1992. Pennsylvania checklist.
Mathis and Bowles, 1992. Missouri distribution, May-September.
Floyd, *et al.,* 1993. South Carolina distribution (Upper Three Runs Creek), July-September.
Moulton, *et al.,* 1993. Texas distribution (Brazos River drainage), July.

DESCRIPTION. Larva: Length 6.0-8.0 mm. Head light yellow, with dark muscle scars on frontoclypeus and parietal sclerites (Figure 10A); muscle scars indistinct in some specimens; antennae short, barely reaching posterior edge of labrum; ventral apotome either triangular and lightly pigmented or indistinct (Figure 10D). Pronotum light brown, with dark muscle scars along lateral and posterior margins (Figure 10B); row of long setae along anterior and lateral margins; transverse, mesal row of 5, long setae on each sclerite. Mesonotum light yellow, with anterior, lateral, and posteromesal muscle scars (Figure 10C); 1 seta present at each anteromesal corner, 7-8 setae along each anterolateral corner; each side with transverse row of 3 posteromesal setae. Metanotum membranous, each side with 1 Sa2 seta and 5-6 Sa3 setae. Metasternum with up to 36 setae. Foretrochantins each with 1 seta. Forelegs each with basal seta on claw; tarsal claws of hind legs longer or equal in length to respective tarsi (Figure 10E). One pair of abdominal gills present, 1 dorsally and 1 ventrally on each of segments II to VIII; tergum IX with 6 large and 4 small setae. **Case:** Length 8.0 mm. Dorsoventrally flattened and composed of sand grains, with larger ballast grains or stones attached laterally (Figure 10F, 10G); tapered and slightly curved.

MATERIAL EXAMINED. DELAWARE: New Castle Co., West Cr. nr. Newark, 1 L, 7 June 1951 (ANSP); **FLORIDA:** Orange Co., Lk. Prima Vista, Orlando, 1 L, 10 June 1992, M. Vogel (CUAC); Orange/Osceola Co., Boggy Cr. @ Rt. 530, 3 L, 12 June 1992, M. Vogel (CUAC); **INDIANA:** Morgan Co., Amur Study Pond, Martinsville Goldfish Hatchery, 1 L, 1 Aug. 1974, A. Provonsha (PU); **ILLINOIS:** Lake Co., Fox Lk., 1 L, 17 May 1938, Ross and Burks (INHS); **MARYLAND:** Frederick Co., Potomac R., betw. Rt. 28 and B & O RR Br., 1 L, 21 Aug. 1975, S. Roback (ANSP); Montgomery Co., Potomac R., 4 mi. below mth. of Monocacy, 1 L, 3 Aug. 1960, S. Roback (ANSP); **NORTH CAROLINA:** Alamance Co., Town Cr. @ Rt. 1200, 3 L, 5 May 1992 (NCDEM); Brunswick Co., Waccamaw R. @ Freeland, 3 L, 7 June 1987 (NCDEM); Columbus Co., Waccamaw R., 2 L, 19 June 1991 (NCDEM); Durham Co., Little R. nr. Orange Factory, 1 L, 12 July 1987 (NCDEM); Harnett Co., Cape Fear R. @ Lillington, 1 L, 12 Sept. 1990 (NCDEM); Johnston Co., Neuse R. @ Clayton, 3 L, 11 July 1986 (NCDEM); Robeson Co., Lumber R., 5 L, 11 Sept. 1991 (NCDEM); Big Marsh Swamp, 3 L, 11 Aug. 1992 (NCDEM); Wilson Co., Contentnea Cr. @ Stationsburg, 2 L, 22 July 1991 (NCDEM); **ONTARIO:** Nipissing D., Algonquin Prov. Pk., Madawaska R., 1 L, 29 June 1973, I. and S. Smith (ROM); str. beside Rt. 11, 3 mi. S Trout Cr., 1 L, 29 Sept. 1972, ROMFP (ROM); **PENNSYLVANIA:** Lancaster Co., Cocalico Cr., SW Reamstown, 1 L, 27 Aug. 1948; J. Richardson (ANSP); Wyoming Co., Susquehana R., 2 mi. S. Meshoppen, 2 L, 16 Aug. 1977, J. Richardson (ANSP); **SOUTH CAROLINA:** Darlington Co., Black Cr. just above Darlington, 2 L, 11 Sept. 1972, S. Roback (ANSP); Oconee Co., L. Hartwell, Coneross Cr. @ end of Durham Brown Rd., 3 L, July 1993, MAF (CUAC); Orangeburg Co., Edisto R. nr. Orangeburg, 1 L, 10 July 1964, S. Roback (ANSP); **TENNESSEE:** Humphreys Co., Tennessee R. (Kentucky Reservoir) nr. New Johnsonville, 1 L, 13 July 1955, S. Roback (ANSP); **TEXAS:** ?, Camp Bullis, farm pond, Lewis Valley Rd., 36 L, 5 Dec. 1960, O. Flint (USNM).

ASSOCIATION, BIOLOGICAL NOTES, AND DISTRIBUTION. Association of larvae and adults for *O. nocturna* is based on rearings from Lake Hartwell, Oconee County, South Carolina (List of Reared Specimens, p. 39). Larvae of this species have been found in a variety of habitats, including streams, rivers, lakes, artificial ponds, and reservoirs. Larvae from lentic environments were found on the undersides of large rocks, cobble, or sticks. Lotic specimens were found upon gravel or sand substrates, sometimes at the base of aquatic vegetation such as water willow, *Justicia americana* (L.) Vahl. Pupae often were found attached to rocks or large, woody debris such as logs or sticks. As reported by many authors, adults of *O. nocturna* often are collected in conjunction with *O. ditissa*. In those habitats where they are collected together, *O. nocturna* is usually more abundant (Haag, *et al.*, 1984, Frazer, *et al.*, 1991, Harris, *et al.*, 1991). *Oecetis nocturna* is known primarily from the eastern United States and Canada (Map 12). Its range includes Alabama, Arkansas, Delaware*, District of Columbia, Florida*, Georgia*, Illinois, Indiana, Kansas, Kentucky, Louisiana, Manitoba*, Maryland*, Missouri, Mississippi, North Carolina, Ohio, Oklahoma, Ontario*, Pennsylvania, Quebec, South Carolina, Tennessee, Texas, Vermont, Virginia, and West Virginia.

DIAGNOSIS. *Oecetis nocturna* is distinguished most readily from other species by its dorsoventrally flattened case with laterally-attached ballast stones.

Oecetis porteri Ross (Figure 11, Map 13)

Oecetis porteri Ross. 1947. Trans. Amer. Entomol. Soc. 73:154-155 (Type locality: Florida, Dade Co., Miami, 20 November 1945, J. Porter).

Gordon, 1984. Florida distribution.

DESCRIPTION. Larva: Length 7.0 mm. Head reddish-brown, with numerous brown muscle scars and many brown reticulations (Figure 11A); antennae reaching to posterior edge of labrum; ventral apotome reddish-brown, subquadrate (Figure 11C). Pronotum reddish-brown, with many brown reticulations (Figure 11B); row of long setae along anterior and lateral margins, transverse, mesal row of 5 setae on each side. Mesonotum reddish-brown, with reticulated pattern of spots; 1 seta present at each anteromesal corner, 4-5 setae present laterally along each side of anterior margin, 2-3 setae on middle of each side. Metanotum membranous, each side with 1 Sa1 seta, 1 Sa2 seta, and 5-7 Sa3 setae. Metasternum with irregular row of 26-28 setae. Coxae of legs with brown reticulations; foretarsi each with 1 basal seta; claws of hindlegs as long as or longer than respective tarsi. Each side of abdomen with 1 dorsal and 1 ventral gill on segments II to VII, pair of pleural gills on segment II; lateral sclerites of segment I brown, each with ventral seta (Figure 11D); tergite IX brown, with 8 large and 6 small setae (Figure 11E); bases of anal prolegs sclerotized, brown. **Case:** Length 8.0 mm. Composed of sand grains, with smooth exterior (Figure (Figure 11E, 11F); tubular, slightly curved, tapered.

MATERIAL EXAMINED. ALABAMA: Covington Co., Lk. Jackson @ Lk. Jackson St. Pk., Florala, 15 L, 14 Mar. 1992, MAF and JBG (CUAC); **FLORIDA:** Baker Co., Ocean Pond @ Olustee Beach Rec. Area, E Lake City, 30 L, 17 Dec. 1991, MAF and J. Moulton (CUAC); same but 2 L, 16 Mar. 1992; Ocean Pond @ U.S. For. Serv. Cpgrd., N side of lake, 30 L, 19 Dec. 1991, MAF (CUAC); Clay Co., Deer Pt. Lk., 2 L, 18 Dec. 1990, L. Burton (CUAC); Hutchinson Lk., off Rt. 214, 1.5 mi. S L. Geneva, 2 L, 15 Mar. 1992, MAF (CUAC); Little Crystal Lk. off Rt. 18 and 100, 2 L, 15 Mar. 1993, MAF (CUAC); Lowry Lk. @ boat ramp, N Keystone Heights, 1 L, 15 Mar. 1992, MAF (CUAC); Orange Co., Lk. Conway, Orlando, 6 L, 11 June 1992, M. Vogel (CUAC); Lk. Down, 1 L, 30 Mar. 1992, M. Vogel (CUAC); Lk. Down, city of Windermere, 1 L, 10 Sept. 1992, M. Vogel (CUAC); Lk. Johns, 1 L, 11 Aug. 1992, M. Vogel (CUAC); Lk. Tibet, 1 L, 10 Sept. 1992; M. Vogel (CUAC); Polk Co., Lk. McCloud, 1 L, 8 Mar. 1951, W. Beck, Jr. (FAM); Spirit Lk. @ Fox Lk., 1 L, 15 May 1936, W. Beck, Jr. (FAM); Union Co., Lk. Butler @ Butler, 1 L, 15 Mar. 1992, MAF (CUAC); **NORTH CAROLINA:** Bladen Co., White Lk. @ White L., 11 L, 27 July 1947, D. Frey (INHS); Columbus Co., Lk. Waccamaw @ Lk. Waccamaw, 20 L, 26 June 1992, MAF and JBG (CUAC); Washington Co., Phelps Lk. off Rt. 1183, S side of lake, 5 L, 23 May 1993, MAF (CUAC); **NOVA SCOTIA:** Big Dam West Lk., Kejimukujik National Park, 1 L, 18 Oct. 1988, B. Bilyj; **SOUTH CAROLINA:** Chesterfield Co., Lk. Robinson, 5 mi. E. Mcbee, 4 L (SCDHEC); Lexington Co., J. B. Martin Study, control pond, 2 L, 28 Mar. 1977 (SCDHEC).

ASSOCIATION, BIOLOGICAL NOTES, AND DISTRIBUTION. The association of larvae and adults for *O. porteri* is based on rearings from Lake Jackson, Covington County, Alabama; Lowry Lake, Clay County, Florida; and Ocean Pond, Baker County, Florida (List of Reared Specimens, p. 39). Within the southeastern United States, *O. porteri* has been found only in sandy-bottomed, natural lakes of the Coastal Plain. At two locations, Lake Jackson, Alabama, and Ocean Pond, Florida, the larvae were collected most abundantly from the bases of emergent grasses (*i.e., Panicum* spp.). However, in Lake Waccamaw, North Carolina, I found larvae attached to the submerged portions of floating-type aquatic vegetation *i.e.,* cow lily, *Nuphar luteum* ssp. *sagittifolium* (Walter) E. O. Beal. In addition to the Coastal Plain of the southeastern United States, *O. porteri* was found to occur in southern Nova Scotia. A single larval specimen was collected from Big Dam West Lake in Kejimukujik National Park. Perhaps this indicates that additional specimens could be found from other localities along the Atlantic Coast of North America. At present, *O. porteri* is known only from four states, Alabama*, Florida, North Carolina*, South Carolina*, and one Canadian province, Nova Scotia* (Map 13).

DIAGNOSIS. *Oecetis porteri* can be distinguished from other *Oecetis* species by its reddish-brown reticulations of the head, pronotum, and mesonotum.

Oecetis pratelia Denning (Map 14)

Oecetis pratelia Denning. 1948. Bull. Brooklyn Entomol. Soc. 43:126 (Type locality: Florida, Hendry Co., LaBelle, 16 July 1939, R. H. Beamer).

Gordon,1984. Florida distribution.

BIOLOGICAL NOTES AND DISTRIBUTION. Nothing is known concerning the biology of *O. pratelia.* The species is known only from one specimen, the holotype, which I have been unable to locate. I also do not know from what possible habitat the adult could have originated. I attempted to find the larva of this species in December 1991 from the area surrounding the type locality, LaBelle, Hendry County, Florida. I collected in the Caloosahatchee River but found no larvae. The river was clear with a sand-gravel bottom and almost no flow. Light-trap collections at this locality produced adults of *O. inconspicua* Complex and *O. osteni.* Morse (1982) listed this species as threatened in Florida, but it actually may be extinct.

Oecetis inconspicua Complex

Seven additional, morphologically distinct, larval forms belonging to the *Oecetis* (*Oecetis*) *furva* Group were discovered and reared to adulthood. I suspected that one or two of these forms would turn out to be those of an unassociated species such as *O. ditissa* or *O. pratelia.* However, the rearings did not produce these results. Instead, the adults of all seven forms appeared to represent "typical" *O. inconspicua* and were morphologically almost identical (except for subtle differences in overall size and genitalic structures, especially the phallus). These findings provide strong support for the suspicion of many researchers that *O. inconspicua* actually represents more than one species. Presently, I have decided to designate these species as *Oecetis* species A-G, all of which are considered to be members of the *O. inconspicua* Complex. Typically, these larval forms have distinct patterns of muscle scars, spots, or other markings on the sclerites of the head and thorax (Figures 12A, 13A, 14A, 15A, 16A). In addition, only the forelegs have a basal seta on the claw, the claws of the hind legs are long and curved (usually at least as long as or longer than its respective tarsus), each foretrochantin has 1 dorsal seta, each lateral sclerite of abdominal segment I is shaped as in Figure 11D and is often darkly pigmented, and dorsal and ventral pairs of gills are present on each of abdominal segments II to VIII, with a single pair of pleural gills on segment II.

Oecetis species A (Figure 12, Map 15)

DESCRIPTION. Larva: Length 7.5 mm. Head with distinct pattern of muscle scars and stripes (Figures 12A, 12C); ventral apotome light yellow, triangular (Figure 12D). Pronotum yellow with dark muscle scars and spots (Figure 12B). Mesonotum light brown, with anterior and posterior, light, brown muscle scars; 1 seta at each anteromesal corner, 7 setae along each anterolateral margin, 3 setae at middle of each side. Metanotum membranous, each side with 1 Sa2 seta, 9 Sa3 setae. Metasternum with irregular row of 32-36 setae. Coxae with 2-4 dark, brown muscle scars. Tergite of segment IX with up to 14 large, 4 small setae along posterior margin. **Case:** Length 8.0 mm. Composed of wood and bits of detritus (Figure 12E) or sand; slightly tapered and curved.

MATERIAL EXAMINED. ALABAMA: Baldwin Co., Turkey Cr. @ Rt. 59, 1 L, 13 Mar. 1992, MAF (CUAC); **NORTH CAROLINA:** Pender Co., Angola Cr., 1 L, May 1993, D. Lenat, NCDEM; **SOUTH CAROLINA:** Aiken

Co., Upper Three Runs Cr. @ Rd. 8-1, Sav. R. Site, 2 L, 2 Apr. 1991, MAF (CUAC); same but 3 L, 9 April; same but 26 Apr.; same but 2 L, 6 Apr. 1993; same but 1 L, 13 Apr. 1979, McEwan and Morse; same but 2 L, 7 Aug. 1979, Kelley and McEwan; same but 1 L, 17 Nov. 1979; same but 1 L, 16 Oct. 1976, Herlong and Prichard; same but 4 Dec.; same but 15 Jan. 1977; same but 13 June 1977.

ASSOCIATION, BIOLOGICAL NOTES, AND DISTRIBUTION. Association of larvae and adults for this species was made through rearings from Upper Three Runs Creek, Aiken County, South Carolina (List of Reared Specimens, p. 39). At this site, the creek is a third order, swift, blackwater stream with rooted vegetation such as *Vallisneria americana* Michaux and *Potamogeton epihydrus* Rafinesque and a channel width of approximately 10 m. The larvae were found on the stream bottom in depositional areas and attached to logs. Those larvae reared in the laboratory constructed the remainder of their case with sand, resulting in a case composed of half detritus and half sand. Larvae of this species were found in three Coastal Plain streams in Alabama, North Carolina, and South Carolina (Map 15). Many of those specimens reported as *O. inconspicua* by Floyd, *et al.* (1993), Harris, *et al.* (1991), and Morse, *et al.* (1980) were probably adults of this species.

DIAGNOSIS. This species can be distinguished from other species within the *O. inconspicua* Complex by the characteristic coloration of its head and pronotum.

Oecetis species B (Figure 13, Map 16)

DESCRIPTION. Larva: Length 7.0 mm. Head with distinct pattern of muscle scars and markings, 2 pale spots along frontoclypeal sutures at middle of frontoclypeus (Figure 13A); ventral apotome light yellow, triangular. Pronotum with distinctive, dark muscle scars and markings (Figure 13B). Mesonotum light brown, with anterior and posterior, dark muscle scars; 1 seta at each anteromesal corner, 12 setae along each anterolateral margin, 3 setae at middle of each side. Mesosternum with 5-6 setae. Metanotum membranous, each side with up to 5 Sa2 setae and 8 Sa3 setae. Metasternum with irregular row of 50-60 setae. Coxae each with 2-4 dark, brown muscle scars. Tergite of segment IX with up to 14 large, 6 small setae along posterior margin. **Case:** Length 8.0 mm. Composition variable, similar to *O. porteri* (Figure 11F) and *Oecetis* sp. A (Figure 12E).

MATERIAL EXAMINED. NORTH CAROLINA: Hyde Co. Lk. Mattamuskeet off Rt. 94, 2 mi. N L. Comfort, 5 L, 23 May 1993, MAF (CUAC); Perquimans Co., Perquimans R. @ Lillin, 1 L, 11 July 1990, D. Lenat (NCDEM).

ASSOCIATION, BIOLOGICAL NOTES, AND DISTRIBUTION. The larva of this species is known only from Lake Mattamuskeet, a clear, sand-bottomed lake in Hyde County, North Carolina and Perquimans River, Perquimans County, North Carolina (Map 16). The larvae were collected along the bottom of Lake Mattamuskeet at the base of emergent grasses.

DIAGNOSIS. This species can be distinguished from other species within the *O. inconspicua* Complex by the distinctive coloration of its head and pronotum.

Oecetis species C (Figure 14, Map 17)

DESCRIPTION. Larva: Length 7.0 mm. Head dark brown, except for pair of pale spots along frontoclypeal sutures at middle of frontoclypeus (Figure 14A); ventral apotome dark brown, subrectangular (Figure 14C). Pronotum dark brown (Figure 14B). Mesonotum dark brown, with 1 seta at each anteromesal corner, 7 setae along each anterolateral margin, 3 setae at middle of each side. Mesosternum with irregular patch of 80-100 setae. Metanotum dark brown, each side with 1-3 Sa2 setae, 8 Sa3 setae. Metasternum with irregular row of 40-50 setae. Dorsum of legs dark brown (Figure 14D); coxae each with 2-4 dark, brown muscle scars. Tergite of segment IX with up to 8 large, 6 small setae along posterior margin. **Case:** Length 9.0 mm. Composed of sand, detritus; slightly dorsoventrally flattened, tapered.

MATERIAL EXAMINED. FLORIDA: Clay Co., Hutchinson Lk., off Rt. 214, 2.5 km S Lk. Geneva, 4 L, 15 Mar. 1992, MAF (CUAC); same but 8 L, 16 Mar. 1992; same but 2 L, 15 Mar. 1993; pond off Duval Rd., Dupont property, 5 km SW Kingsley Lk., 8 L, 15 Mar. 1993, MAF and JBG (CUAC); **SOUTH CAROLINA:** Chesterfield Co., Bee Lk. Sandhills National Wildlife Refuge, 1 L, 5 Dec. 1981, E. Younginer (SCDHEC).

ASSOCIATION, BIOLOGICAL NOTES, AND DISTRIBUTION. Association of larvae and adults of this species was based on rearings from two small ponds (~ 0.5 ha) in Clay County, Florida (List of Reared Specimens, p. 39). Each of the two ponds were clear, shallow, and had a firm, sand substrate. Larvae in each locality were found along the pond bottom, usually at the base of emergent grasses. Some of the larvae seemed to be partially buried in the organically enriched sand surrounding the roots. The roots of the plants had to be shaken violently in order to dislodge the larvae. This species is known from three localities in Florida and South Carolina (Map 17).

DIAGNOSIS. This species can be distinguished from other species within the *O. inconspicua* Complex by its dark brown head and pronotum, 80-100 setae on the mesosternum, and dark coloration of the dorsal surface of the hind legs.

Oecetis species D (Figure 15, Map 18)

DESCRIPTION. Larva: Length 5.0-6.0 mm. Head and thorax pale (Figures 15A, 15B); mesonotum with 1 setaat each anteromesal corner, 4 setae along each anterolateral margin, 3 setae at middle of each side. Metanotum membranous, each side with 1 Sa2 seta, 4 Sa3 setae. Metasternum with regular, transverse row of up to 28 setae. Tergite of segment IX pale, with up to 6 large, 4 small setae along posterior margin. **Case:** Length 6.5-7.0 mm. Identical to that of *O. nocturna* (Figure 10F).

MATERIAL EXAMINED. NORTH CAROLINA: Columbus Co., Lk. Waccamaw @ L. Waccamaw, 25 L, 26 June 1992, MAF and JBG (CUAC).

ASSOCIATION, BIOLOGICAL NOTES, AND DISTRIBUTION. Larval and adult association for this species was based on rearings from Lake Waccamaw, a clear, shallow (< 5 mm), sand-bottomed lake in Columbus County, North Carolina (List of Reared Specimens, p. 39). Lake Waccamaw represents the only known locality for this larva (Map 18). The larvae were found along with *O. porteri* at the base of cow lily, *Nuphar luteum* ssp. *sagittifolium* (Walter) E. O. Beal. The case of this species is virtually identical to that of *O. nocturna*.

DIAGNOSIS. This species can be distinguished from other species within the *O. inconspicua* Complex by its pale coloration and dorsoventrally flattened case.

Oecetis species E (Figure 16, Map 19)

DESCRIPTION. Larva: Length 8.0 mm. Head, pronotum, and mesonotum with distinct pattern of dark stripes (Figure 16A, 16B); ventral apotome dark brown, nearly triangular (Figure 16C); mesonotum with 1 seta at each anteromesal corner, 3-5 setae along each anterolateral margin, 3 setae at middle of each side. Metanotum membranous, each side with 1 Sa2 seta, 2 Sa3 setae. Metasternum with regular row of 4-7 setae. Coxae of legs each with 2-4 dark, lateral muscle scars. Tergite of segment IX with up to 6 large, 4 small setae along posterior margin. **Case:** Length 10.5 mm. Identical to that of *Oecetis* sp. A (Figure 12E) or composed of spirally arranged pieces of vegetation (Figure 24E).

MATERIAL EXAMINED. SOUTH CAROLINA: Barnwell Co., Castor Bay, Sav. R. Site, 50 L, 6 Apr. 1993. MAF and D. Bidlack (CUAC); same but Craig's Pond, 6 L; Horry Co., Bear Swamp @ Rt. 110, 1 Sept. 1994, H. Gayman and R. Renfrow (SCDHEC).

ASSOCIATION, BIOLOGICAL NOTES, AND DISTRIBUTION. Association of larvae and adults for this species was based on rearings from Castor Bay and Craig's Pond, two Carolina Bay lakes on the Savannah River Site, Barnwell County, South Carolina (List of Reared Specimens, p. 39). Larvae were found among emergent grasses at both sites (Map 19). A larval specimen from Horry County, South Carolina had constructed a very unusual case composed of spirally-arranged plant pieces.

DIAGNOSIS. This species can be distinguished from other species within the *O. inconspicua* Complex by the distinctive coloration of its head and pronotum and the small number of setae (4-7) on the metasternum.

Oecetis species F (Figure 17, Map 20)

DESCRIPTION. Larva: Length 6.0-6.5 mm. Head and pronotum with distinct pattern of muscle scars, spots, and markings (Figure 17A, 17B); ventral apotome subrectangular, very pale except dark anterior edge. Mesonotum with 1 seta at each anteromesal corner, 4 setae along each anterolateral margin, 3 setae at middle of each side. Metanotum membranous, each side with 1 Sa2 seta, 4 Sa3 setae. Metasternum with regular, transverse row of 14-16 setae. Coxae of legs each with 2-4 dark, lateral muscle scars. Tergite of segment IX with up to 10 large, 4 small setae along posterior margin. **Case:** Length 7.5 mm. Similar to that of *O. porteri* (Figure 11F), exterior not as smooth.

MATERIAL EXAMINED. FLORIDA: Osceola Co., L. Tohopekaliga @ Southport Pk., off Rt. 531, 28 L, May 1988, R. Butler (CUAC); same but 14 L, 16 Mar. 1993, MAF and JBG; **SOUTH CAROLINA:** Lexington Co., J. B. Martin Study, control pond, 8 L, 28 Mar. 1977 (SCDHEC).

ASSOCIATION, BIOLOGICAL NOTES, AND DISTRIBUTION. Larval and adult association for this species was based on rearings from Lake Tohopekaliga, Osceola County, Florida (List of Reared Specimens, p. 39). The larvae were collected around several species of aquatic macrophytes that were growing along shore near a boat ramp. Other caddisfly larvae found at this locality included *O. cinerascens*, *O. osteni*, and another leptocerid species, *Triaenodes furcella* Ross.

DIAGNOSIS. This species can be distinguished from other species within the *O. inconspicua* Complex by the distinctive coloration of its head and pronotum.

Oecetis species G (Map 21)

DESCRIPTION. Larva: Length 7.5-8.0 mm. Head, pronotum, mesonotum pale, with no distinct markings (Figure 15A, 15B). Metanotum membranous, each side with 1 Sa2 seta, 6 Sa3 setae. Metasternum with row of 45-50 setae. Tergite of segment IX pale, with 6 large, 4 small setae along posterior margin. **Case:** Length 9.5-10.0 mm. Composed of sand and bits of rock, tubular in shape.

MATERIAL EXAMINED. ARIZONA: Pima Co., cow tank @ King Anvil Ranch, off Rt. 286, many L, 28 Feb. 1993, J. K. Moulton (CUAC).

ASSOCIATION, BIOLOGICAL NOTES, AND DISTRIBUTION. Association of the adult and larva of this species was based on rearings from a small (50 m in diameter), shallow (1 m), elliptical, temporary pond in Pima County, Arizona. The pond, or cow tank, served as a water supply for cattle of the King Anvil Ranch. Because of constant visitation by cattle, the pond's water remained very turbid. The larvae were found in abundance along the pond's bottom.

DIAGNOSIS. This species can be distinguished from other species within the *O. inconspicua* Complex by the pale color of its head and pronotum.

Oecetis (*Oecetis*) *ochracea* Group

Oecetis ochracea (Curtis) (Figure 18, Map 22)

Leptocerus ochraceus Curtis. 1825. British Entomol. 2-15: Pl. 57. (Type locality: Britain).

Edwards, 1966. Tennessee distribution (Reelfoot Lake), 8 May.
Hilsenhoff, *et al.,* 1972. Wisconsin distribution (Pine-Popple River), 15 July-12 August.
Longridge and Hilsenhoff, 1973. Wisconsin distribution, 15 July-12 August.
Newell and Potter, 1973. Montana distribution, July.
Harris, *et al.,* 1980. North Dakota distribution, June-August.
Nimmo and Scudder, 1978. British Columbia distribution, 7-8 July.

Etnier and Schuster, 1979. Tennessee distribution, 8 May.
Parker and Voshell, 1981. Virginia checklist.
Flannagan and Flannagan, 1982. Manitoba distribution.
Roemhild, 1982. Montana distribution and ecological notes.
Ruiter and Lavigne, 1985. Wyoming distribution, 8 July-27 August
Herrman, *et al.,* 1986. Colorado distribution, 27 July-20 August.
Garono and MacLean, 1988. Ohio distribution (wetlands).
Jones and Resh, 1988. Montana distribution.

DESCRIPTION. Larva: Length 13-14 mm. Head yellow, with dark muscle scars on frontoclypeus and parietal sclerites (Figure 18A); 3 large spots on dorsum of head identical to those of *O. immobilis*; antennae reaching posterior margin of labrum; ventral apotome dark, trapezoidal (Figure 18C). Pronotum yellow, with darker anterior area (Figure 18B); 4 muscle scars present on each side along posterior margin of dark area; additional muscle scars present along posterolateral margins; row of long setae along anterior and lateral margins, transverse row of 5-6 setae near middle of each side. Mesonotum yellow, with light brown markings; 1 seta present at each anteromesal corner, approximately 25 setae present at each anterolateral corner, 3 setae present mesally on each side. Mesosternum with row of 4-6 setae. Metanotum membranous, each side with 1 Sa2 seta and 14-18 Sa3 setae. Metasternum with irregular row of 50-90 setae. Foretrochantins each with 1 dorsal seta. Claws of forelegs each with basal seta; coxae of legs each with 2-4 lateral, small, dark muscle scars; hind tarsi longer than respective claws. Dorsal and ventral abdominal gills present on segments II to VIII, pleural gills on segment II; tergite IX with 26-30 setae, all approximately equal in size. **Case:** Length 15.0 mm. Slightly tapered and composed of sand grains (Figure 18D); smooth to rough exterior, often curved, fragile compared to other *Oecetis* cases.

MATERIAL EXAMINED. ALASKA: ponds nr. Kayigyalik Lk., Yukon Delta, 1 L, Aug.-Sept. 1985, J. Maciolek (USNM); **ALBERTA:** Elkwater Lk., Cypress Hill Prov. Pk., 1 L, 26 June 1962, GBW (ROM); **MINNESOTA:** Mahnomen Co., pond on Rt. 113, Waubun, 1 L, 27 Aug. 1970, GBW (ROM); **NORTHWEST TERRITORY:** pond on S side of Dempster Hwy., km 503, 2 L, 30 July 1980, ROMFP (ROM); same but 1 L, 7 July 1979; **SASKATCHEWAN:** Margo Lk., southern Sask., 1 L, 10 July 1938 (INHS); Montreal L., 2 L, Summer 1955, D. Dawson (ROM); Murray Lk., 8 L, 15 June 1962, GBW (ROM); **YUKON:** Cadzow Lk., 20 mi. E Old Crow, 1 L, 14 July 1981, R. Jaagumagi (ROM); along shoreline, Haines Rd. @ Dezadeash Lk. Cpgrd., 2 L, 19 June 1980, ROMFP (ROM).

ASSOCIATION, BIOLOGICAL NOTES, AND DISTRIBUTION. Descriptions and keys for *O. ochracea* have been provided by Ross (1944), Lepneva (1966), and Wallace, *et al.* (1990). According to Lepneva (1966), this species frequents lakes, ponds, and reservoirs, usually on solid or slightly silted bottoms. Roemhild (1982) reported this species inhabiting warm-water lakes in Montana. It may occur several meters deep and is often an early colonizer of new reservoirs and lakes (Wallace, *et al.,* 1990). In North America, this Holarctic species is known primarily from the northern United States and Canada as follows: Alaska, Alberta, British Columbia, California, Colorado, Manitoba, Minnesota, Montana, North Dakota, Northwest Territory*, Ohio, Ontario*, Saskatchewan, South Dakota, Wisconsin, Wyoming, and Yukon*. However, there also have been adult collections made from three southern states: North Carolina*, Tennessee, and Virginia (Map 22).

DIAGNOSIS. Larvae of *O. ochracea* can be identified by three large head spots in combination with a transverse row of 50-90 setae on the metasternum, one seta on each foretrochantin, and 26-30 setae along the posterior margin of tergum IX.

Oecetis (Oecetis) lacustris Group

Oecetis immobilis (Hagen) (Figure 19, Map 23)

Setodes immobilis Hagen. 1861. Syn. Neur. N. Amer. p. 283 (Type locality: Canada).

Blickle, 1966. Maine distribution, 19 July-6 August.
Corbet, *et al.,* 1966. Quebec distribution (St. Helen's Isl., Montreal), 9 July-2 September.
Nimmo, 1966. Quebec distribution.

Longridge and Hilsenhoff, 1973. Wisconsin distribution, 13-26 August.
Newell and Potter, 1973. Montana distribution, July-August.
Hilsenhoff, 1975. Wisconsin checklist, keys to genera.
Roy and Harper, 1975. Quebec distribution, 13 July-2 September.
Masteller and Flint, 1979. Pennsylvania distribution (Lake Erie), 9 August.
Roy and Harper, 1979. Quebec distribution, 1 July-2 September.
Wimmer, 1979. Vermont distribution, 11-19 August.
Harris, *et al.,* 1980. North Dakota distribution, June-August.
Flannagan and Flannagan, 1982. Manitoba distribution.
Roemhild, 1982. Montana distribution, notes on ecology.
Huryn and Foote, 1983. Ohio distribution, 17 June-10 September.
Resh, *et al.,* 1983. Manitoba distribution (Southern Indian Lake).
Waltz and McCafferty, 1983a. Indiana distribution, 26 June-30 August.
Richardson and Clifford, 1986. Alberta distribution (Sturgeon R.), ecology, and phenology.
Garono and MacLean, 1988. Ohio distribution (wetlands).

DESCRIPTION. Larva: Length 6.0 mm. Head yellow, with dark muscle scars on frontoclypeus and parietal sclerites (Figure 19A); 1 large, dark spot present on each postgenal parietal between eye and frontoclypeal suture, with additional spot present at posterior end of frontoclypeus; antennae reaching past posterior edge of labrum; ventral apotome dark, short, 4 times as wide as long (Figure 19D). Pronotum yellow, with many muscle scars along posterior margin (Figure 19B); row of long setae along anterior and posterior margins, transverse row of 6 setae on middle of each side. Mesonotum yellow, with dark muscle scars and broad, U-shaped, dark band (Figure 19C); 1 seta at each anteromesal corner, 6-9 setae at each anterolateral corner, 3 setae on middle of each side. Metanotum membranous, each side with 1 Sa2 seta and 5-6 Sa3 setae. Metasternum with regular row of up to 26 setae. Foretrochantins each with 10-11 setae. Forelegs each with basal seta on claws; tarsi of hindlegs greater than or equal to length of respective claws. Each side of abdomen with 1 dorsal and 1 ventral gill on segments II to VIII, pair of pleural gills on segment II; lateral sclerite of segment I yellow, each with 1 ventral seta; posterior margin of tergum IX with 6 large and 4 small setae. **Case:** Length 6.0-7.0 mm. Tubular, composed of sand grains (Figure 19E); tapered, slightly curved, with smooth exterior.

MATERIAL EXAMINED. ALBERTA: Sturgeon R. below Lac St. Anne, 9 L, 1981-1982, H. Clifford and J. Richardson (CUAC); same but 18 L, 15 June 1981, J. Richardson (ROM); **ILLINOIS:** Lake Co., Fox Lk. @ Fox Lk., 6 L, 15 May 1936, Ross and Mohr (INHS); E Fox Lk. @ Mineola, 1 L, 9 June 1938, Mohr and Burks (INHS); Channel Lk., 1 L, 31 May 1936, H. Ross (INHS); **OHIO:** Ottawa Co., Put-in-Bay, 1 L, 17 Aug. 1937, Mohr (INHS); **ONTARIO:** Lk. Ontario, Prince Edward Bay, 2 L, Aug. 1967 (ROM); Norfolk Co., Long Pt., Gravelly Bay, 3 L, 20 July 1963, TY (ROM).

ASSOCIATION, BIOLOGICAL NOTES, AND DISTRIBUTION. Ross (1944) collected and described two distinctive, but unassociated, larval *Oecetis* species, one of which he designated as *Oecetis* species A. The head of this larva had a very recognizable pattern of three spots that formed the corners of a triangle (Figure 812, Ross 1944). Ross suspected that this was the larva of *O. immobilis* for two reasons. Adult specimens of *O. immobilis* had been collected from the same localities as species A, and the larval structures of species A were very similar to those of *O. inconspicua*, a closely related species based on adult structures. Ross' suspicion was confirmed by the association of larvae and adults by Richardson and Clifford (1986), a study of the ecology and phenology of several Trichoptera in the Sturgeon River, Alberta. This species has been collected from both lakes and streams. Richardson and Clifford (1986) reported that *O. immobilis* was univoltine in Alberta (Sturgeon R.) and found in a variety of microhabitats, macrophytes, cobble, and woody debris. Ross (1944) stated that in Illinois *O. immobilis* was taken abundantly only around the glacial lakes of the northeast corner of the state. *Oecetis immobilis* has been collected only in the northern United States and Canada (Map 23). Records for this species include the following: Alberta, British Columbia, Illinois, Indiana, Maine, Manitoba, Michigan, Minnesota, Montana, New Brunswick, New York, North Dakota, New Hampshire, New York, Nova Scotia, Ohio, Ontario, Pennsylvania, Quebec, Saskatchewan, Vermont, Wisconsin, and Yukon.

DIAGNOSIS. *O. immobilis* can be distinguished from other *Oecetis* species by its distinctive ventral apotome and three dark spots on the head.

OECETIS (*PSEUDOSETODES*) ULMER 1905

Oecetis (*Pseudosetodes*) *avara* Group

Oecetis avara (Banks) (Figs. 20, 21, Map 24)

Setodes avara Banks. 1895. Trans. Amer. Entomol. Soc. 22: 316 (Type locality: Sherbrooke, Canada, June, P. A. Bégin).

Blickle and Morse, 1966. Maine distribution, 8 July-6 August.
Unzicker, *et al.,* 1970. Arkansas distribution, April-October.
Edwards, 1973. Texas distribution, 17 Mar.-July 6.
Resh, 1975. Kentucky distribution, 24 June-28 August.
Roy and Harper, 1975. Quebec distribution, 1 June-4 August.
Etnier and Schuster, 1979. Tennessee distribution, 17 June-27 September.
Neves, 1979. Massachusetts distribution, June-August.
Roy and Harper, 1979. Quebec distribution, 1 June-12 August.
Wimmer, 1979. Vermont distribution, 1 July.
Mingo and Gibbs, 1980. Maine distribution (Narraguagus River), 28 June-31 July.
Smith and Lehmkuhl, 1980. Idaho, Montana, Ontario, and Saskatchewan distributions; larval, pupal, and adult descriptions.
Parker and Voshell, 1981. Virginia checklist.
Holzenthal, *al.,* 1982. Mississippi distribution and abundance, 10 June-20 September.
Roemhild, 1982. Montana, distributional and ecological notes.
Unzicker, *et al.,* 1982. North and South Carolina distribution, July-mid-August.
Hamilton, *et al.,* 1983. Kansas distribution.
Harris, *et al.,* 1983. Alabama distribution (Tyro Creek watershed), June.
Huryn and Foote, 1983. Ohio distribution, 4 July-10 September.
Waltz and McCafferty, 1983a. Indiana distribution, 29 May-10 September.
Waltz and McCafferty, 1983b. New Mexico distribution.
Cobb, *et al.,* 1984. Manitoba distribution (Cowan Creek).
Harris, *et al.,* 1984. Alabama distribution (Cahaba River system), May-September.
Lake, 1984. Delaware distribution, June-August.
Steven and Hilsenhoff, 1984. Wisconsin distribution (Otter Creek), notes on biology.
Ruiter and Lavigne, 1985. Wyoming distribution, 12 June-30 August.
Bilger, 1986. New England checklist (Maine, Massachusetts, New York, and Vermont).
Bowles and Mathis, 1989. Arkansas distribution (mountainous regions), April-October.
Morse, *et al.,* 1989. South Carolina distribution (Upper Three Runs Creek).
Glover and Tarter, 1990. West Virginia distribution, May-September.
Harris and Lago, 1990. Alabama distribution, May-October.
Tarter. 1990. West Virginia distribution, May-September.
Harris, *et al.,* 1991. Alabama distribution, May-October.
Bowles and Mathis, 1992. Oklahoma distribution, June-September.
Masteller and Flint, 1992. Pennsylvania checklist.
Mathis and Bowles, 1992. Missouri distribution, May-September.
Floyd, *et al.,* 1993. South Carolina distribution (Upper Three Runs Creek); May; June; August.
Moulton, *et al.,* 1993. Texas distribution (Brazos River drainage), March-November.
Moulton, *et al.,* 1994. Arizona distribution, April-June.

DESCRIPTION. Larva: Length 6.5-9.0 mm. Dorsum of head light yellow with contrasting dark muscle scars on postgenal parietals and frontoclypeus (Figure 20A), muscle scars lighter in color or indistinct in some specimens; parietals with lateral, darkened anterior and posterior areas (Figure 20C); posterior area with 3-4 dark muscle scars; subocular line indistinct; antennae short, reaching anteriorly to posterior edge of labrum; ventral apotome much wider than long, with rounded lateral margins and pair of small, posterolateral, subtriangular sclerites (Figure 20D). Pronotum yellow, with light, transverse band posteriorly (Figure 20B); row of many long setae along anterior and lateral margins; each side with transverse row of 6-7 setae posterior to anterior row. Mesonotum light yellow, with 1 seta near each anteromesal corner; as many as 19 setae on each anterolateral area, three setae on middle of sclerite. Metanotum membranous, each side with single seta present on Sa2 and Sa3 areas. Trochantin with up to 7 setae. Tarsus of hindleg longer than claw (Figure 21); tarsal claws of legs each with basal seta (Figures 22 E, 22F). Dorsal hump of abdomen with two groups of micro-hooks or spines on the dorsal hump of abdominal segment I (Figure 20E); abdominal gills present dorsally and ventrally on segment II to VII; pleural gills present on segment II; tergite of segment IX with up to 6 large and 4 small setae on posterior margin. Claws of anal prolegs each with one accessory hook. **Case:** Length. 9.0 mm. Composed of sand grains, with rough exterior; curved in early instars; similar to that of *O. disjuncta* (Figures 22G, 22H).

MATERIAL EXAMINED. ALBERTA: Pembina R. Prov. Pk., Rt. 16, 1 L, 18 June 1962, GBW (ROM); Oldman R. @ Monarch, 1 L, 25 June 1962, GBW (ROM); **ARIZONA:** Apache Co., E. Fk. Black R. @ Aspen, Apache Nat. Forest Cpgrd., 5 L, 4 May 1981, McCafferty, Provonsha, and Bloodgood (PU); Coconino Co., Oak Cr. @ Sedona, 1 L, 29 Apr-2 May 1991, GBW (ROM); Oak Cr. @ Red Rock Crossing Cpgrd., 1 L, 19 Oct. 1966, GBW, TY, and Odum; Yavapai Co., Verde R., 18 mi. N Jerome, 1 L, 3 July 1966, GBW (ROM); same but @ Rt. 179 nr. Camp Verde, 8 L, 4 July 1966; **CALIFORNIA:** Trinity Co., Trinity R. @ US 299, .5 mi. W Del Loma, 1 L, 21 June 1985, ROMFP (ROM); **COLORADO:** Garfield Co., Colorado R. @ Rifle Railroad Station, 1 L, 27 June 1977, D. E. Ruiter; **ILLINOIS:** Jo Daviess Co., Apple R. Canyon St. Pk., 2 L, 27 June 1940, Mohr and Riegel (INHS); same but 13 L, 6 June 1940, Mohr and Burks; Jo Daviess Co., Galena R., Council Hill, 5 L, 5 June 1940, Mohr and Burks (INHS); **MAINE:** Kennebec Co., Kennebec R. @ Pishon Ferry, 1 L, 15 Aug. 1975, J. Richardson (ANSP); **MANITOBA:** Roseau R., 58 L, 27 May 1981, Schefter, Fuller, and Flannagan (ROM); **MARYLAND:** Montgomery Co., Potomac R., 2 mi. below mth., Monocacy, 2 L, 20-21 Aug. 1957, S. Roback (ANSP); same but @ White's Ferry, 1 L, 4 Aug. 1960; **MINNESOTA:** Blue Earth Co., Blue Earth R., Rapidan, 2 L, 11 Oct. 1937, Mohr (INHS); Clearwater Co., Mississippi R. @ Itasca St. Pk., 9 L, 1 June 1981, Schefter and Fuller (ROM); Mississippi R., 5 mi. N L. Itasca Post Office, 1 L, 28 July-19 Aug. GBW; Ottertail Co., Ottertail R. @ mi. .57, Broken Down Dam Rd., 5 L, 21 Sep. 1979, N. Kirsch (UM); Pine Co., Snake R., 7 L, 29 May 1977, G. Daussin (UM); Renville Co., Minnesota R. @ mi. 243.2, 1 L, N. Pothoff (UM); **MISSISSIPPI:** Simpson Co., Strong R. @ Rt. 28, 2.3 mi. W Pinola, 1 L, 1 July 1979, R. Holzenthal, (CUAC); **NEW YORK:** Delaware Co., Beaver Kill R., Roscoe, Twin Island Cpgrd., 1 L, 11 June 1988, Schefter and MacCulloch (ROM); Oneida Co., Sconondoa Cr. nr. Vernon, 1 L, 11 July 1952, S. Roback (ANSP); same but 4 L; same but 1 L, 12 July 1952; **NORTH CAROLINA:** Burke Co., Linville R., 9 L, Aug. 1983, D. Lenat (NCDEM); same but 3 L, 27 July 1986; Alleghany Co., New R. @ Amelia, 5 L, 25 July 1988, D. Lenat (NCDEM); Rutherford Co., Cove Cr. @ Lk. Lure, 4 L, 21 July 1986, D. Lenat (NCDEM); **NOVA SCOTIA:** Queens D., Westfield R., 2 L, 1 June 1981, M. Dadswell (ROM); **OHIO:** ?, Auglaize R., 3.5 mi. W Kalida, above jct. with Ottawa R., 1 L, 6 July 1956, S. Roback (ANSP); same but 5 July 1955; **OKLAHOMA:** Johnston Co., Pennington Cr., S Reagan, 1 L, 10 May 1970, GBW and TY (ROM); Murray Co., Honey Cr., Turner Falls Pk., Rt. 77 S Davis, Arbuckle Mts., 2 L, 7-8 May 1970, GBW and TY (ROM); same but effluent str. from Arbuckle Dam @ Rt. 110 N Dougherty, 5 L, 10 May 1970; **ONTARIO:** Durham D., Wilmont Cr. @ Newcastle, 2 L, 10 July 1952, GBW (ROM); N Ganaraska R., crossing nr. Canton, 9 L, 19 June 1973, GBW and TY; same but crossing 5th line W Perrytown, 1 L, 19 June 1973; same but 2km S Garden Hill, 1 L, 22 June 1986, Vineyard and Phillips; Halton D., Credit R., Glen Williams, 2 L, 16 July 1952, GBW (ROM); same but 15 Sept. 1952; same but 1L; Credit R. @ Norval, 3 L, 30 May 1973, TY and Frania; Hastings D., Moira R., Latta, 15 L, 25 May 1973, TY and H. Frania (ROM); same but 7 L, 25 May 1976, TY and Avruch; same but 1 L, 15 Oct. 1968, GBW, TY, and Barr; same but 1 L, 2 July 1976, TY; same but 9 L, 20 June 1972, ROMFP; same but 1 L, 27 June 1971, GBW; same but 1 L, 25 July 1971, TY; same but 36 L, 24 June 1973, ROMFP; same but 4 L, 13 June 1976, TY; same but Moira R. @ Rt. 101, 8 L, 24 June 1973, ROMFP; same but Moira R. @ Rt. 401, 2 L, 15 Aug. 1976, TY; same but 3 L, 13 June 1976; same but Moira R. @ Fry's Landing Prk., 4 L, May 1979, Schefter and Schefter; same but Skootamatta R. @ Rt. 7, SW Kaladar, 1 L, 26 Aug. 1971, ROMFP; Huron D., Maitland R., 9.8 mi. W Listowel, Rt. 86, 3 L, 17 June 1976, TY and Avruch (ROM); small str. @ Rt. 12, 2.3 mi. S Winthrop, 9 L, 16 June 1976; Lanark D., Mississippi R. @ Carleton Place, Rt. 29, 1 L, 26 Aug. 1971, ROMFP (ROM); Northumberland D., small trib. Ganaraska R., 2 L, 26 May 1976, TY and Avruch (ROM); Peel D., Credit R. @ Cheltenham, 2 L, 16 July 1952, GBW (ROM); same but @ Streetsville, 3 L, 12 June 1952; same but 11 L, 25 Aug. 1952; same but 2 L, 30 May 1973, GBW and TY; same but. @ Erindale, 1 L, 11 Aug. 1976, TY; same but 2 L, 30 May 1973, TY and Frania; same but @ Churchville, 4 L, 18 Sept. 1952, GBW; same but 14 L, 12 June 1952; Perth D., Thames R., below St. Mary's, 1 L, 14 June 1968, J. Richardson (ANSP); same but 7 L, 12 June 1968; Thunder Bay D., Sturgeon R. @ Rt. 11, 11.1 mi. NE Jellicoe, 1 L, 22 June 1971, ROMFP (ROM); **OREGON:** Douglas Co., Cow Cr. @ Riddle, 3 L, 9 June 1975, N. Anderson (ROM); Josephine Co., Rogue R. nr. Galice, 1 L, 22 June 1985, JCM (CUAC); **QUEBEC:** Rivière du Nord, St. Andre Est., 1 L, 5 Aug. 1981, P. Catling (ROM); **SOUTH CAROLINA:** Oconee Co., Little R. @ Rt. 24, 1 L, 25 Apr. 1991, MAF (CUAC); **TEXAS:** ?, Guadalupe R. @ Sequin, 1 L, 8 Sept. 1950, T. Dolan (ANSP);

VIRGINIA: Wythe Co., Reed Cr., 3 L, 20 Dec. 1984 Schefter and Vineyard (ROM); **WISCONSIN:** Florence Co., Pine R., end of 2nd Rd. S jct. Co. Rd. N & D, 6 L, 4 June 1981, Schefter and Fuller (ROM).

ASSOCIATION, BIOLOGICAL NOTES, AND DISTRIBUTION. Association of larvae and adults is based on rearings accomplished by Ross (1944) and Smith and Lehmkuhl (1980). *Oecetis avara* is strictly a lotic species occurring in streams and rivers of varying size, current speed, and turbidity. In South Carolina, for example, this species occurs in clear, rocky streams (*e.g.,* Little River, Chattooga River) of the upper Piedmont and Mountains, as well as blackwater streams (*e.g.,* Upper Three Runs Creek) of the Sandhills and Coastal Plain. Many authors (Edwards, 1973; Holzenthal, *et al.,* 1982; Waltz and McCafferty, 1983a) have noted that *O. avara* is most abundant in larger streams and rivers. In North America as a whole, the adults have been collected from April to October. Smith and Lehmkuhl (1980) suggested that this species may be bivoltine in the warm waters of the Saskatchewan River system. Larvae of *O. avara* have been collected most frequently while clinging to the surfaces of rocks or woody debris. Specimens of *O. avara* from Wisconsin were decidedly carnivorous, as their gut contents were 100% animal remains by volume (Shapas and Hilsenhoff, 1976). I examined one specimen of *O. avara* from Ontario that had partially consumed a larval chironomid, *Synorthocladius* sp. Based on larval and adult specimens, *O. avara* is known from Alabama, Alberta*, Arizona, Arkansas, California*, Colorado, Delaware, District of Columbia, Georgia, Idaho, Illinois, Indiana, Kansas, Kentucky, Maine, Manitoba, Maryland, Michigan, Minnesota, Mississippi, Missouri, New York, North Carolina, North Dakota, New Hampshire, New Jersey, New Mexico, Nova Scotia, Ohio, Oklahoma, Ontario, Pennsylvania, Quebec, Saskatchewan, South Carolina, Tennessee, Texas, Utah, Vermont, Virginia, West Virginia, and Wisconsin (Map 24). Because of their morphological similarities, many of the literature records for this species and *O. disjuncta* (Banks) are suspect and possibly incorrect (Smith and Lehmkuhl, 1980). Therefore, literature records published before 1980 are omitted from the publication list above. However, most records from the eastern and southeastern United States are included, because only *O. avara* has been found to exist in these regions.

DIAGNOSIS. With the exception of *O. disjuncta, O. avara* is distinguishable from all other North American species by its two groups of micro-hooks or spines on the dorsal hump of abdominal segment I. It differs from *O. disjuncta* in its shorter and wider ventral apotome and lighter head coloration.

Oecetis disjuncta (Banks) (Figure 22, Map 25)

Oecetina disjuncta Banks. 1920. Bull. Mus. Comp. Zool. Harvard 64: 351 (Type locality: California: Arroyo Seco Canyon, San Gabriel Mountains, 17 June, Switzer's Camp, F. Grinnell).

Smith and Lehmkuhl, 1980. California, Michigan, Oregon, Saskatchewan, South Dakota, and Utah distributions; larval, pupal, and adult descriptions.
Roemhild, 1982. Montana distribution and ecological notes.
Waltz and McCafferty, 1983b. New Mexico distribution, May.
Ruiter and Lavigne, 1985. Wyoming distribution, 19 June-29 August.
Herrmann, *et al.,* 1986. Colorado distribution, 20 June-1 September.

DESCRIPTION. Larva: Length 7.0-8.0 mm. Head brown to dark yellow, with dark muscle scars on frontoclypeus and parietal sclerites (Figure 22A); parietals with darkened anterior and posterior lateral areas (Figure 22C); posterior area with 3-4 dark muscle scars; additional muscle scar groups present posterior and ventral to darkened areas; subocular line indistinct; antennae short, reaching anteriorly to posterior edge of labrum; ventral apotome brown, 2 times as wide as long, with rounded lateral margins and pair of small, posterolateral, subtriangular sclerites (Figure 22D). Pronotum with light, anterior, transverse band (Figure 22B); brown, transverse area in middle, covering about two-thirds of the length of the pronotum; pale or light area posteriorly, with about 7 muscle scars; row of many long setae along anterior and lateral margins; transverse row of 7-8 setae posterior to anterior row on each side. Mesonotum brown anteriorly, pale posteriorly; 1 seta near each anteromesal corner; as many as 19 setae on each anterolateral area, three setae on middle of each sclerite. Metanotum membranous, each side with 1-2 Sa2 setae and 1-3 Sa3 setae. Each foretrochantin with 3-6 dorsal setae. Tarsus of each midleg and hindleg longer than claw (Figure 22F); tarsal claws of all legs each with basal seta (Figure 22E, F). Dorsal hump of abdomen with two groups of micro-hooks or spines on the dorsal hump of abdominal segment I (Figure 20E); abdominal gills present dorsally and ventrally on segment II to VII; pair of pleural gills present on segment II; abdominal segment IX with up to 6 large and 4 small setae on posterior margin. **Case:** Identical to that of *O. avara* (Figures 22G, 22H).

MATERIAL EXAMINED. ALBERTA: Cypress Hills Prov. Pk., 30 km E Elkwater, @ Battle Cr. Cpgrd., 1 L, 18 Aug. 1981, C. Parker (ROM); Sturgeon R. @ Lac St. Anne, 1 L, 5 May 1981, J. Richardson (ROM); **ARIZONA:** Cochise Co., Cave Cr., SW Res. Stat., 5 mi. W Portal, 1 L, 26 June 1966, GBW (ROM); Coconino Co., W Frk. Oak Cr., 3 L, 5 Dec. 1986, GBW (ROM); same but 3 L, 3-4 Dec. 1986; Oak Cr. @ Sedona, 1 L, 4 Dec. 1986, GBW (ROM); same but 5 Dec. 1986; Oak Cr. @ Encinosa Picnic Grd., 1 L, 5 July 1967, TY (ROM); same but Banjo Bill Cpgrd., 4 L, 19 Oct. 1966, GBW, TY, and Odum; Yavapai Co., Deception Gulch nr. Jerome, 1 L, 29 July 1984, Unzicker and Sanderson (INHS); **CALIFORNIA:** Los Angeles Co., Millard Cr. in Millard Canyon, N Pasadena, 6 L, 8 Dec. 1981, GBW (ROM); str. in Arroyo Seco Canyon, Switzer's Camp N Pasadena, 20 L, 8 Dec. 1981, GBW (ROM); Monterey Co., seepage str., Molera St. Pk., 8 L, D. Nye (ROM); same but 7 L, 25 Feb. 1975; spring str. beside Big Sur R., 2 L, 7 Jan. 1975, D. Nye (ROM); San Diego Co., str. in Green Valley Cpgrd., Cuyamaca Rancho St. Pk., 1 L, 3 Dec. 1981, GBW (ROM); San Luis Obispo Co., Santa Rita Cr., Cypress Mtn., 23 L, 21 Dec. 1974, (ROM); Shasta Co., Hat Cr., Fall R. Mills, 3 L, 30 June 1969, C. Apperson (ROM); Stanislaus Co., Adobe Cr., 12 mi. W Patterson, 1 L, 25 Apr. 1948, H. Leech (INHS); Trinity Co., Weaver Cr. nr. Douglas City, 1 L, 8 Oct. 1976, J. Furnish (INHS); **COLORADO:** Fremont Co., Badger Cr., about 100 yards upstream from the Arkansas R. confluence, 1 L, 19 Oct. 1988, D. E. Ruiter; **IDAHO:** Caribou Co., Blackfoot R., 9 L, 5 Sept. 1973, R. Newell (ROM); **MANITOBA:** Riding Mtn. Nat. Pk., Edwards Cr., 2 L, 11 June 1962, GBW (ROM); **MICHIGAN:** Crawford Co., AuSable R., Lovell, 81 L & pupae, 22 May 1936, Frison and Ross (INHS); Emmet Co., W Br. Maple R. @ Rt. 31, Pellstou, 2 L, 13 June 1972, Barr and Smith (ROM); **NEW MEXICO:** Catron Co., W Frk. Gila R., nr. Heart Bar Wild. Area, 2 L, 3 Sept. 1977, J. Wojtowicz (UT); W Fk. Gila R. @ Scorpion Cpgrd., Gila Nat. For., 1 L, 5-6 May 1981, McCafferty, Provonsha, and Bloodgood (PU); Eddy Co., Lincoln Nat. Forest, Sitting Bull Falls, 4 L, 4 July 1988, R. Vineyard (ROM); Mora Co., Coyote Cr. @ Rt. 434, mi. marker 20, 3 L, 24 May 1991, MAF (CUAC); Mora Co., str. 6.4 mi. S Mora @ Murphy Lake Rd., 3 L, 25 May 1991, MAF (CUAC); **OREGON:** Harney Co., Blitzen R. @ Paige Springs, 4 L, 3 july 1981, G. Courtney (ROM); same but Trout Cr., 1 L, 27 Aug. 1983, R. Wisseman (ROM); **SASKATCHEWAN:** Cold R. @ Cold L., 2 L, 30 Aug. 1976, D. Smith (ROM); Englishman R. @ Rt. 26, 4 L, 2 May 1977, D. Smith (ROM); Goodsoil, mossy str. on Rt. 224, 4.5 mi. N Meadow L. Prov. Pk., 6 L, 16 Aug. 1970, ROMFP (ROM); Loon L., Makwa R., 5 mi. N Rt. 26, 1 L, 16 Aug. 1970, ROMFP (ROM); **UTAH:** Wasatch Co., 3 mi. N Heber City, small str. nr. Provo R. @ Rt. 189, 4 L, 29 Apr. 1981, J. Weaver (CUAC); Wasatch Co., Provo R. @ River Rd., 9 L, 24 Apr. 1981, C. Parker (ROM); str. @ jct. Rt. 189 and Rd. to Wallsberg, 8 L, 30 Apr. 1981, Schefter and Schuster (ROM); Wasatch/Summit Co., Provo R. on Rt. 189, 1 L, 26 Apr. 1981, C. Parker (ROM); Washington Co., Zion Nat. Prk., str. below Weeping Rock, 1 L, 1 July 1978, GBW (ROM); Santa Clara R. @ Pine Valley, 1 L, 26 May 1971, McCafferty (PU); **WISCONSIN:** Sauk Co., Otter Cr., 5 L, 17 Sept. 1974, C. Gorsuch (CUAC); **WYOMING:** Teton Co., Firehole R., upper Geyser Basin, Yellowstone Nat. Pk., 1 L, 11 July 1961, GBW (ROM).

ASSOCIATION, BIOLOGICAL NOTES, AND DISTRIBUTION. Association of larvae and adults for this species was accomplished through rearings made by Smith and Lehmkuhl (1980). Although the adults of this species have been reported from both lakes and streams (Roemhild, 1982; Herrman, *et al.,*1986), larval records are only available from lotic habitats. The larvae seem to prefer fast-flowing, rocky streams, some of which occur at altitudes of over 2,700 m (Herrman, *et al.,* 1986). In Saskatchewan, Smith and Lehmkuhl (1980) reported that this species was univoltine, with overwintering larvae and a peak emergence period that was earlier than that of *O. avara*. Throughout its range, the adults of this species were collected on dates ranging from early June to early September. *Oecetis disjuncta* is distributed mainly across western North America but is found as far east as Michigan and Ontario (Map 25). Records for this species include the following: Alberta*, Arizona*, California, Colorado, Idaho*, Manitoba*, Michigan, Montana, New Mexico, Ontario*, Oregon, Saskatchewan, South Dakota, Utah, Wisconsin*, and Wyoming*.

DIAGNOSIS. Excluding *O. avara*, the rows of micro-hooks or spines on the dorsal hump of abdominal segment I clearly separate this species from other North American species. Larvae of *O. disjuncta* can be separated from those of *O. avara* by their darker head coloration and longer, more narrow, ventral apotome.

Oecetis (*Pseudosetodes*) *lais* Group

Oecetis osteni Milne (Figure 23, Map 26)

Oecetis osteni Milne. 1934. Stud. N. Am. Trich. 1: 17,19 (Type locality: New York, Alexandria Bay, August, O. Sacken).

Blickle and Morse, 1966. Maine distribution, 8-31 July.
Corbet, *et al.,* 1966. Quebec distribution (St. Helen's Island, Montreal).
Longridge and Hilsenhoff, 1973. Wisconsin distribution, 13-17 August.
McElravy, *et al.,* 1977. Ohio distribution.
Neves, 1979. Massachusetts distribution, June-September.
Morse, *et al.,* 1980. South Carolina distribution (Upper Three Runs Creek), March; May-June; August-September.
Harris, *et al.,* 1982. Florida distribution (Eglin Air Force Base), 11 May; 19 September.
Holzenthal, *et al.,* 1982. Louisiana and Mississippi distribution, 5 April-18 October.
Unzicker, *et al.,* 1982. North and South Carolina distribution, late March-mid-September.
Harris, *et al.,* 1983. Alabama distribution (Tyro Creek watershed), July.
Huryn and Foote, 1983. Ohio distribution, 24 July-15 August.
Waltz and McCafferty, 1983a. Indiana distribution, 26 May-28 July.
Harris, *et al.,* 1984. Alabama distribution (Cahaba River system), September-October.
Garono and MacLean, 1988. Ohio distribution (wetlands).
Harris and Lago, 1990. Alabama distribution, March-October.
Frazer, *et al.,* 1991. Alabama distribution (Little River drainage), June; August.
Harris, *et al.,* 1991. Alabama distribution, March-October.
Masteller and Flint, 1992. Pennsylvania checklist.
Floyd, *et al.,* 1993. South Carolina distribution (Upper Three Runs Creek), July-October.

DESCRIPTION. Larva: Length 6.0 mm. Head yellow, with dark muscle scars on frontoclypeus and parietal sclerites (Figure 23A); row of 4 muscle scars along each side of coronal suture, with adjacent, parallel row of 3-4 muscle scars; antennae long, reaching to middle of labrum; mandibles with deep crease running posteriorly from apical tooth (Figure 23E); ventral apotome subquadrate to subtriangular, with pair of large, posterolateral, triangular sclerites (Figure 23D). Pronotum yellow, with pale area along posterior margin (Figure 23B); dark muscle scars present on posterior half; row of long setae on anterior and lateral margins; transverse row of 5 setae on middle of each side, with dark spot surrounding each setal base. Mesonotum with large, irregular, brown area each side of meson (Figure 23C); 1 seta present at each anteromesal corner, 4-6 setae along anterolateral margins of each side, 3-4 setae at middle of each brown area. Metanotum membranous, each side with 1 Sa2 seta and 1 Sa3 seta. Metasternum with up to 3 pairs of setae. Tarsal claws of hind legs longer than or equal to length of respective tarsi; legs each with basal seta on claw. Abdomen with dorsal and ventral gills on segments II to IV, pleural gills absent; tergum of segment IX with 6 large and 4 small setae. **Case:** Length 6.0 mm. Tubular-shaped, constructed of sand grains, with smooth exterior (Figure 23F); often tapered and curved in early instars.

MATERIAL EXAMINED. ALABAMA: Covington Co., Lk. Jackson @ Lk. Jackson St. Pk., Florala, 1 L, 14 Mar. 1992, MAF and JBG (CUAC); **FLORIDA:** Gadsden Co., Rocky Comfort Cr. @ Rt. 12, 1 L, 14 Jan. 1987, M. Pescador (FAM); Orange Co., Lk. George, 1 L, 17 Aug. 1992, M. Vogel (CUAC); Osceola Co., Lk. Tohopekaliga @ Southport Park, off Rt. 531, 5 L, 16 Mar. 1993, MAF and JBG (CUAC); Union Co., Lk. Butler @ Butler, 1 L, 15 Mar. 1992, MAF (CUAC); Washington Co., Pate Lk., off Rt. 279, 14 L, 18 Mar. 1993, MAF (CUAC); **ILLINOIS:** Lake Co., Antioch, 1 L, 18 May 1938, Ross and Burks (INHS); Channel Lk. nr. Antioch L., 1 L, 18 May 1938, Ross and Burks (INHS); **MINNESOTA:** Clearwater Co., Itasca Lk., Itasca St. Pk., 27 L, 14 Aug. 1972, GBW (ROM); **NEW YORK:** Tompkins Co., Dryden Lk. @ Dryden, 1 L, 15 July 1958, O. Flint (USNM); **ONTARIO:** Kenora D., Lodge Lk., Rt. 17, 7.4 mi. NW English R., 10 L, 18 June 1971, ROMFP (ROM); Foley Lk., 49.2 km E Sioux Lookout on Rt. 516, 11 L, 18 June 1984, R. Jaagumagi and J. Prideaux (ROM); Leeds Co., Chaffeys Locks, Lk. Opinicon, 1 L, 28 June 1972, I. Smith (ROM); Muskoka D., small bay off Lk. Muskoka, 2 mi. S Bala on Rt. 69; 1 L, 10 June 1975, GBW and TY (ROM); Echo L., Baysville, 2 L, 20 July 1966, I. Smith (ROM); Thunder Bay D., small lake, 1.9 km N Rt. 11 on Rd., 25 km W jct. Rt. 11, 1 L, 4 July 1984, ROMFP (ROM); small lake beside Rt. 17, 2 mi. W. Rossport, 2 L, 7 June 1971, ROMFP (ROM); **SOUTH CAROLINA:** Barnwell Co., Craig's Pond, Savannah R. Site, 47 L, 6 Apr. 1993, MAF (CUAC); same but 7 L, 14 Mar. 1993; Sumter Co., Cane Savannah Cr., 1 mi.

S Cane Savannah on St. Paul's Church Rd., 1 L, 2 Sept. 1986, P. Grant (CUAC); 2nd Mill Pond and W Liberty St., Sumter, 3 L, 28 Oct. 1986, P. Grant (CUAC).

ASSOCIATION, BIOLOGICAL NOTES, AND DISTRIBUTION. Association of larvae and adults for *O. osteni* was based on rearings from Lake Jackson, Covington County, Alabama and Lake Tohopekaliga, Osceola County, Florida (List of Reared Specimens, p. 39). Ross (1944) described this species as *Oecetis* species B, based on one specimen from Channel Lake, Lake County, Illinois. He suspected that it was the larva of *O. osteni*, but difficulties in rearing prevented him from making the association. *Oecetis osteni* is primarily a lentic species but can be found in slow portions of streams and rivers. The larvae have been collected most abundantly from natural lakes, specifically in association with aquatic vegetation. I found specimens clinging to water primrose, *Ludwigia peploides* (HBK) Raven ssp. *peploides*, and emergent grasses, *Panicum* spp. Pupal cases were found always attached to vegetation. The range of *Oecetis osteni* includes one band from Minnesota and western Ontario to New Brunswick and another from Louisiana to North Carolina (Map 26). Between these regions, it is absent entirely from three states, Kentucky, Tennessee, and West Virginia. Distributional records for this species include the following states and provinces: Alabama, Florida, Illinois, Indiana, Louisiana, Maine, Massachusetts, Michigan, Minnesota, Mississippi, New Brunswick, New Hampshire, New Jersey, New York, North Carolina, Ohio, Ontario, Pennsylvania, Quebec, Rhode Island*, South Carolina, Virginia, and Wisconsin.

DIAGNOSIS. *O. osteni* can be distinguished from other *Oecetis* species by the distinctive row of muscle scars along the coronal suture, the creased mandibles, and the irregular, brown areas of the mesonotum.

Oecetis (*Pseudosetodes*) *parva* Group

Oecetis daytona Ross (Map 27)

Oecetis daytona Ross. 1947. Trans. Am. Entomol. Soc. 73:153-154. (Type locality: Florida, Volusia Co., Daytona Beach, 27 July 1945, G. T. Riegel).

Holzenthal, *et al.,* 1982. Mississippi distribution, 24 May; 5 August.
Unzicker, *et al.,* 1982. North and South Carolina distribution, early July.
Gordon, 1984. Florida distribution.
Harris, 1990. Alabama rare and endangered invertebrates.
Harris and Lago, 1990. Alabama distribution, April-August.
Harris, *et al.,* 1991. Alabama distribution, April-August.

BIOLOGICAL NOTES AND DISTRIBUTION. The larvae of this species may occur in lentic or lotic habitats, because it has been collected around both lakes and streams. The adults have been collected irregularly and in small quantities (two to three specimens) from April to September. Both the female and larva of this species remain to be associated. The larva of a closely related species, *O. pratti* Denning, has been described by Flint (1968) and was found crawling over rocks and leaves in small mountain streams of the Dominican Republic. *Oecetis daytona* appears to be endemic to the Southeast, specifically the Coastal Plain (Map 27). Harris, *et al.* (1991) collected adults of *O. daytona* in Alabama along small, sand and gravel bottom streams of the Southern Pine Hills. Only one specimen has been found in South Carolina, approximately 10 miles south of Columbia. Besides these two states, it also has been collected in Florida (Alachua, Columbia, Duval, Lee, and Volusia Counties) and Mississippi (George and Perry Counties). Due to its rare occurrence, *Oecetis daytona* has been listed as rare or threatened in Alabama (Harris,1990) and Florida (Morse,1982).

Oecetis parva (Banks) (Figure 24, Map 28)

Setodina parva Banks. 1907. Proc. Entomol. Soc. Wash. 8: 130 (Type locality: Florida, Osceola Co., Kissimmee, November).

Gordon, 1984. Florida distribution.
Harris, *et al.,* 1991. Alabama distribution (Wright's Creek), May.

DESCRIPTION. Larva: Length 3.0-3.5 mm. Head light yellow, with few, faint muscle scars on parietal sclerites (Figure 24A); some darkening along frontoclypeal sutures mesad to eyes and adjacent to coronal suture; antennae long, reaching anterior margin of labrum; ventral apotome light yellow, nearly circular, with dark anterior stripe, surrounded by pale area (Figure 24C); mandibles each with 3 teeth, basal tooth blunt with median indentation (Figure 24D). Pronotum light yellow, with row of long setae along anterior and lateral edges (Figure 24B); transverse row of 5-6 setae each side of meson; transverse, faint, darker stripe present across middle of sclerite. Mesonotum light yellow, with 1 long seta at each anteromesal corner, 2 setae on middle of each side. Mesosternum with 10-12 setae. Each side of metanotum with 1 Sa2 seta, 0 Sa3 setae. Metasternum without setae. Tibiae and tarsi of hindlegs each with constriction at middle; hind claws at least as long as respective tarsi; tarsal claws of all legs each with 1 basal seta. Abdomen with dorsal and ventral gills on segments II to IV, pleural gills absent; tergite of segment IX with 6 large, 4 small setae. **Case:** Length 4.0 mm. Composition variable, including short, transverse segments of vegetation arranged spirally (Figure 24E), long, sharp, linear plant pieces placed irregularly (Figure 24F), flat, quadrate, plant pieces, or mixtures of all three.

MATERIAL EXAMINED. FLORIDA: Washington Co., Lucas Lk. @ public boat ramp, 9 km SSE Vernon, 4 L, MAF and JBG (CUAC).

ASSOCIATION, BIOLOGICAL NOTES, AND DISTRIBUTION. Association of larvae and adults for this species was based on rearings from Lucas Lake, Washington County, Florida. A total of only seven larvae was collected from this locality. Three specimens were chosen for rearing (List of Reared Specimens, p. 39). Larvae of this species were found attached to milfoil, *Myriophylum laxum* Shuttlw. ex Chapm., which was growing adjacent to a boat ramp. Although only seven specimens were collected, many empty cases were found, indicating that this species may be fairly common at this site. This rare southeastern endemic has been collected only from two states, Alabama and Florida (Map 28). In Alabama, one male of *O. parva* was reported by Harris, *et al.* (1991) from Wright's Creek, a small, sand-bottomed stream of the Coastal Plain. Florida records include the counties of Alachua (near Gainesville), Osceola (at Kissimmee), and Washington (Lucas Lake).

DIAGNOSIS. This species can be distinguished from other *Oecetis* species based on its small size, long antennae, blunt basal tooth of the mandible, divided tibiae and tarsi of hind legs, peculiar case construction, and limited distribution.

KEY TO KNOWN LARVAE OF THE GENUS *OECETIS* IN NORTH AMERICA[1,2,3]

1. Dorsal hump of abdominal segment I with 4-6 rows of micro-hooks on each side (Fig. 20E) *Oecetis (Pseudosetodes) avara* Group **2**
1'. Dorsal hump of abdominal segment I without micro-hooks **3**

2(1). Head either yellow, with dark muscle scars (Figs. 20A, 20C) or paler yellow, with indistinct muscle scars; ventral apotome approximately four times wider than long (Fig. 20D); common in streams throughout North America .... ***O.* (*Ps.*) *avara* (Banks), p. 27**
2'. Head brown to dark yellow, with darker muscle scars (Figs. 22A, 22C); ventral apotome approximately twice as wide as long (Fig. 22D); mainly streams of western North America .... ***O.* (*Ps.*) *disjuncta* (Banks), p. 29**

3(1'). Postgenal sclerites separated by distinctive brown and pale areas (Fig. 7C); lateral hump of abdominal segment I sometimes with dark, elongate sclerite (Fig. 7E); *Oecetis* (*Quaria*) *scala* Group **4**
3' Postgenal sclerites not separated by dark and pale areas; shape of lateral sclerite of abdominal segment I variable **5**

4(3). Ventral apotome trapezoidal, with small posterolateral, triangular sclerites (Fig. 7D); lateral hump of first abdominal segment with dark, elongate, lateral sclerite (Fig. 7E), sclerite with 1 ventral seta; metasternum with 12 setae; streams of southeastern United States .... ***O.* (*Q.*) *morsei* Bueno-Soria, p. 13/*sphyra* Ross, p. 15**
4'. Ventral apotome subrectangular, four times wider than long, with small posterolateral, triangular sclerites (Fig. 8C); lateral sclerite of abdominal segment I light yellow, with 2 ventral setae; metasternum with 2 setae; streams, rivers of central and northeastern United States .... ***O.* (*Q.*) *eddlestoni* Ross, p. 13**

5(3'). Metasternum with 2 setae; swimming hairs present on hind tibiae (Fig. 3G); hind tarsi longer than tarsal claws (Fig. 3G); case constructed of root or plant fragments placed transversely (Figs. 4A, 4B, 5D); *Oecetis* (*Pleurograpta*) **6**
5'. Metasternum with more than 2 setae, often with row of 10 or more; swimming hairs absent on hind tibia (Fig. 10E); hind tarsi as short as or shorter than tarsal claws (Fig. 10E); case constructed of sand grains or irregularly-placed plant or detrital pieces (Figs. 10F, 10G, 11F) **8**

6(5). Head brown with light muscle scars (Figs. 3A, 3C); case constructed either of short twigs or roots (angled "log cabin" appearance) (Fig. 4A) or thin, flat, quadrate, plant fragments (Fig. 4B); on vegetation in lakes, some slow-moving streams of eastern North America .... ***O.* (*Pl.*) *cinerascens* (Hagen), p. 6**
6'. Head yellow with dark muscle scars or pale (Figs. 5A, 6A); case exterior more rounded and often curved, especially in early instars (Fig. 5D); on roots, woody debris, and vegetation in streams of eastern North America **7**

7(6'). Head with scattered, well-defined muscle scars (Fig. 6A); on root mats, woody debris of lotic habitats in eastern North America .... ***O.* (*Pl.*) *persimilis* (Banks), p. 10**
7'. Head without dark, well-defined muscle scars (Fig. 5A); on root mats, woody debris of lotic habitats in southeastern North America .... ***O.* (*Pl.*) *georgia* Ross, p. 8**

8(5'). Mesonotum with pair of irregular dark areas each side of meson (Fig. 23C); coronal suture bordered by row of 3-4 dark muscle scars on each side (Fig. 23A); left mandible blunt with deep crease running from apical tooth (Fig. 23E); case constructed of sand grains (Fig. 23F) .... ***O.* (*Ps.*) *osteni* Milne, p. 31**
8'. Mesonotum without distinct pair of irregular dark areas; coronal suture bordered by 0-2 muscle scars; left mandible without crease; case constructed of plant material or sand grains **9**

9(8'). Antennae long, reaching at least to anterior edge of labrum (Figs. 9A, 24A) **10**
9'. Antennae short, reaching only to posterior edge of labrum (Figs. 10A, 11A) **11**

10(9). Head with light, faint muscle scars (Fig. 24A); ventral apotome without posterolateral sclerites (Fig. 24C); basal tooth of mandible tooth blunt (Fig. 24D); case diverse, constructed of plant materials (Figs. 24E, 24F); on vegetation in natural lakes of Florida .... ***O.* (*Ps.*) *parva* (Banks), p. 32**

10'. Head light brown with dark muscle scars (Fig. 9A); ventral apotome with posterolateral sclerites (Fig. 9C); basal tooth of mandible sharp; case constructed of mixture of plant and detrital material (Fig. 9D); streams in Arizona and Colorado .... ***O.* (*Q.*) *arizonica* Denning, p. 16**

11(9'). Dorsum of head with large, dark spot present on each postgenal parietal between eye and frontoclypeal suture, with additional spot present at posterior end of frontoclypeus (Figs. 18A, 19A) **12**

11'. Dorsum of head either pale with no distinct markings or with other combinations of dark muscle scars, spots, or stripes; ***Oecetis* (*O.*) *furva* Group** **13**

12(11). Metasternum with row of 50-90 setae; ventral apotome trapezoidal, roughly two times as wide as long (Fig. 18C); coxae of legs with 2-4 muscle scars; tergite IX with 26-30 setae; mainly lentic, northern North America .... ***O.* (*O.*) *ochracea* (Curtis), p. 24**

12'. Metasternum with row of about 26 setae; ventral apotome subrectangular, short, five times as wide as long (Fig. 19D); coxae of legs without muscle scars; tergite IX with approximately 10 setae; primarily lakes, but also slow rivers, streams of northern North America .... ***O.* (*O.*) *immobilis* (Hagen), p. 25**

13(11'). Head and pronotum usually with several light brown muscle scars (Fig. 10A); case constructed of sand or rock pieces with larger ballast stones attached to the sides (Figs. 10F, 10G); lotic and lentic habitats of eastern North America .... ***O.* (*O.*) *nocturna* Ross, p. 18**

13'. Head and pronotum pale or with various combinations of dark muscle scars, spots, or stripes (Figs. 11A, 11B, 12A, 12B, 13A, 13B, 14A, 14B); case constructed of variety of materials; if case with ballast stones, then each side of metanotum with fewer than 5 Sa3 setae and metasternum with fewer than 28 setae; wide spread **14**

14(13'). Head and pronotum covered with dense, brown markings in addition to well-defined muscle scars (Figs. 11A); ventral apotome reddish-brown, subrectangular (Fig. 11C); case constructed of sand, curved, and with a very smooth exterior (Fig. 11F); sandy-bottomed lakes of Alabama, Florida, and North Carolina .... ***O.* (*O.*) *porteri* Ross, p. 20**

14'. Head and pronotum with variable markings (Figs. 12A, 12B, 13A, 13B, 15A, 15B); ventral apotome variable (Figs. 12D, 14C, 16C); case constructed of sand or plant/detrital pieces; diverse habitats; *O. (O.) inconspicua* Complex[3] **15**

15(14'). Head and pronotum pale (Fig. 15A, 15B) **16**

15'. Head and pronotum with dark muscle scars, stripes, or bars (Figs. 12A, 12B, 13A, 13B, 14A, 14B, 16A, 16B, 17A, 17B) **17**

16(15). Metasternum with row of at least 45 setae; case tubular (similar to Fig. 18) .... ***O.* (*O.*) species G, p. 24**

16'. Metasternum with row of 28 setae or less; case dorsoventrally flattened, with ballast stones attached laterally (Fig. 10F) .... ***O.* (*O.*) species D, p. 23**

17(15'). Mesosternum with irregular patch of 80 to 100 setae; case somewhat flattened dorsoventrally, composed of sand; dorsum of hind legs dark brown (Fig.14D) .... ***O.* (*O.*) species C, p. 22**

17'. Mesosternum with no more than 6 setae; case tubular, composed of sand (Fig. 11F) or woody debris (Fig. 12E); dorsum of hind legs same color as rest of leg **18**

18(17'). Head with dark longitudinal stripe on middle of frontoclypeus (Fig. 16A); ventral apotome dark brown (Fig. 16C) .... ***O.* (*O.*) species E, p. 23**

18'. Head without dark longitudinal stripe on frontoclypeus (Figs. 12A, 13A, 17A); ventral apotome light brown or yellow (Fig. 12D) **19**

19(18'). Metanotum with no more than 4 Sa3 setae; metasternum with row of 14-16 setae ... ***O.* (*O.*) species F, p. 24**

19'. Metanotum with 8 to 9 setae; metasternum with over 30 setae **20**

20(19'). Pronotum light brown (Fig. 13B); metasternum with row of at least 50 setae .... **O. (O.) species B, p. 22**
20'. Pronotum yellow, with dark muscle scars posteriorly; metasternum with row of up to 36 setae .... **O. (O.) species A, p. 21**

---

[1]Key is designed for use with mature larvae (fourth or fifth instar).

[2]Larvae of *O.* (*Ps.*) *daytona* Ross, *O.* (*O.*) *ditissa* Ross, *O.* (*Q.*) *ouachita* Moulton, *O. (Q). ozarkiensis* Moulton, *O. (O.) pratelia* Denning, and *O.* (*O.*) *scala* Milne have not been associated.

[3]The actual number of species belonging to the *O.* (*O.*) *inconspicua* Complex is unknown. It contains at least seven species, *Oecetis* species A-G, that were discovered during the course of this study.

NORTH AMERICAN *OECETIS* SPECIES: PHYLOGENETIC CONSIDERATIONS

Based on adult morphology, Chen (1993) divided the genus *Oecetis* into four monophyletic subgenera, *Oecetis* MacLachlan, *Pleurograpta* Wallengren, *Pseudosetodes* Ulmer, and *Quaria* Milne, all of which have representatives in North America. He also established 21 species groups and constructed phylogenies to reflect their historical relationships. He suggested that larval data may provide additional characters and thereby make his phylogeny more comprehensive. Although *Oecetis* larvae are fairly uniform in their morphology, a number of larval characters were found that provide additional support for monophyly in one subgenus (*Pleurograpta*) and several species groups. Conversely, no evidence was found supporting alternative hypotheses.

The subgenus *Pleurograpta* has five species for which larval descriptions are available. These include three North American species, *O. cinerascens*, *O. georgia*, and *O. persimilis*, described in this study and two Palearctic species, *notata* (Rambur) and *testacea* (Curtis), described by Wallace (1981). The hind tibiae of all five species uniquely share the presence of two rows of long setae, or swimming hairs, which provide an oar-like function enabling the larva to swim.

Larval characters were found that also support several species groups established by Chen (1993). The North American *O. scala* Group can be distinguished from other species groups by the division of the lateral portion of the postgenal parietal sclerites into distinct, yellow-brown and pale areas (key couplet 3). Likewise, the two associated members of the *O. avara* Group, *O. avara* and *O. disjuncta*, uniquely share the presence of four to six rows of micro-hooks on each side of the dorsal hump of abdominal segment I (key couplet 1). The only associated members of the *O. parva* Group, *O. parva* and *O. pratti,* share an unusual constriction or division of the hind tibiae and tarsi. Finally, within the *O. lacustris* Group, which includes the associated species, *O. immobilis* and *O. lacustris* Pictet, the ventral apotome is almost identical (Fig. 4.19D and Fig. 810B; Lepneva, 1966), and there are at least 10 setae present on the foretrochantin.

Although the use of immature forms in making phylogenetic inferences has been debated (Wiggins, 1981; Schmid, 1979), there appear to be instances when knowledge of larval forms has led to the development of more accurate and improved phylogenies. In addition, Resh (1976) found that species groups derived from larval characteristics of Trichoptera were sometimes congruent with those derived from adults. At present, this also appears to be true for *Oecetis*. There is considerable larval evidence supporting previously established species groupings. As more larvae are described in the future, additional larval characters can be studied to see if there is further support for and congruence with the phylogenies proposed by Chen (1993).

CONCLUSIONS

The caddisfly genus *Oecetis* is the largest genus in the Leptoceridae, with over 280 species world-wide (Chen, 1993). At least 29 species of *Oecetis* occur in America north of Mexico, and associations have been made for 22 of these species. The actual number of North American species remains unknown due to uncertainties in the *O. inconspicua* Complex. At least seven species are known that are morphologically and behaviorly distinct as larvae, but resemble "typical" *O. inconspicua* as adults. A total of seven is probably a very conservative estimate, because I have examined many other distinct, but morphologically similar, preserved specimens that also could be included in this complex. Both the actual number of species occurring in the Complex and the larval identity of *O. inconspicua* remain to be discovered. A more thorough understanding of this complex will precede any attempt at a complete synopsis of the taxonomy and biology of *Oecetis* species in North America. Larvae of seven North American species, *O. daytona*, *O. ditissa*, *O. inconspicua*, *O. ouachita*, *O. ozarkiensis*, *O. pratelia*, and *O. scala*, remain to be associated with their corresponding adults. Larval characters found in this study provide additional support for some of the phylogenetic hypotheses of Chen (1993), who relied on adult characters.

Larvae of the genus *Oecetis* are unique among the Leptoceridae because they are strictly predaceous and have developed elongate, cutting-type mouthparts in association with this behavior. In addition, their maxillary palpi extend far beyond the anterior edge of the labrum, and they have numerous secondary setae on the labrum.

Oecetis larvae are found in a variety of aquatic habitats, including natural lakes, reservoirs, artificial ponds, bogs, springs, small streams, and large rivers. Lentic species tend to frequent shallow areas containing aquatic vegetation, or they are substrate dwellers, some of which may occur at depths of over 10 m. Lotic species may be attached to rocks and cobble in the stream middle, or they may occur along stream margins where they inhabit submerged roots of riparian vegetation, other woody debris, or aquatic macrophytes. Several species that are found more commonly in lentic situations also may occur on vegetation in slow portions of streams or rivers. Adult specimens are collected frequently in coastal areas, indicating that some species may inhabit brackish water.

Case construction is highly variable between species. Cases may be constructed of fine sand grains or rock pieces or plant materials such as leaves, woody debris, or short segments of stems, twigs, or roots placed transversely. The case often is tapered and curved, especially in early instars. A few cases are dorsoventrally flattened with larger ballast stones attached to the sides.

In America north of Mexico, the majority of *Oecetis* species are eastern in distribution. Six species, *O. daytona*, *O. georgia*, *O. morsei*, *O. parva*, *O. pratelia*, and *O. sphyra*, are endemic to the Southeast. Only one species, *O. arizonica* Denning is endemic to the west (Arizona and Colorado). Other species with both eastern and western distributions include *O. disjuncta*, *O. immobilis,* and *O. ochracea.*

LIST OF REARED SPECIMENS

Oecetis cinerascens (Hagen)
FLORIDA: Osceola Co., L. Tohopekaliga @ Southport Pk., off Rt. 531, 1 male pupa; Polk Co., Reedy Lake @ boat ramp nr. Frostproof, 1 male pupa; **SOUTH CAROLINA:** Pickens Co., Willards pond @ Clemson Univ. dairy farm, 1 male pupa.

Oecetis georgia Ross
FLORIDA: Duval Co., trib. of Six Mile Cr. @ Rt. 17, 1 male (INHS); **SOUTH CAROLINA:** Aiken Co., Upper Three Runs Cr. @ Co. Rt. 114, 2 female pupae, 2 male pupae (ROM); Upper Three Runs Cr. @ U.S. 278, 1 female pupa (USNM).

Oecetis morsei Bueno-Soria
SOUTH CAROLINA: Aiken Co., Upper Three Runs Cr. @ Rd. 8-1, Sav. R. Site, 1 female (INHS); Upper Three Runs Cr. @ U. S. 278, 1 female pupa (ROM), 1 male (USNM).

Oecetis nocturna Ross
SOUTH CAROLINA: Oconee Co., L. Hartwell, Coneross Cr. @ end of Durham Brown Rd., 2 females (INHS), 2 males (ROM).

Oecetis osteni Milne
ALABAMA: Covington Co., L. Jackson @ L. Jackson St. Pk., Florala, 1 male pupa (INHS); **FLORIDA:** Osceola Co., L. Tohopekaliga @ Southport Pk., off Rt. 531, 1 female pupa (ROM), 2 males (USNM).

Oecetis parva (Banks)
FLORIDA: Washington Co., Lucas L. @ boat ramp, 9km SSE Vernon, 1 female pupa (INHS), 1 male pupa (ROM), 1 male pupa (USNM).

Oecetis persimilis (Banks)
SOUTH CAROLINA: Pickens Co., Rocky Bottom Cr. @ Co. Rt. 100, 1 male pupa (INHS); Oconee Co., Chattooga R. @ Rt. 28, 2 female pupae, 2 male pupae, 1 female (ROM); Little R. @ Co. Rt. 24, 1 female pupa, 2 male pupae (USNM); Spartanburg Co., Pacolet R. @ Co. Rt. 31, 1 female pupa, 1 male (CUAC).

Oecetis porteri Ross
ALABAMA: Covington Co., L. Jackson @ L. Jackson St. Pk., Florala, 3 females, 1 male (INHS); **FLORIDA:** Baker Co., Ocean Pond @ Olustee Beach Rec. Area, E Lake City, 2 males (ROM); Clay Co., Lowry L. @ boat ramp, N Keystone Hghts, 1 female, 1 male (USNM); **NORTH CAROLINA:** Columbus Co., L. Waccamaw @ L. Waccamaw, 1 male pupa, 2 males (CUAC).

Oecetis sphyra Ross
SOUTH CAROLINA: Spartanburg Co., Pacolet R. @ Co. Rt. 31, 2 female pupae (CUAC), 1 male pupa (INHS), 1 female (ROM), 2 males (USNM).

Oecetis species A
SOUTH CAROLINA: Upper Three Runs Cr. @ Rd. 8-1, Sav. R. Site, 1 female pupa, 1 female, 1 male pupa (INHS).

Oecetis species B
NORTH CAROLINA: Hyde Co., L. Mattamuskeet, off Rt. 94, 2 mi. N Lake Comfort, 1 male (ROM).

Oecetis species C
FLORIDA: Clay Co., Hutchinson L., off Co. Rt. 214, 2.5km S Lake Geneva, 1 male pupa, 1 female, 1 male (USNM); pond on Dupont property, off Duval Rd., 5km SW Kingsley L., 1 male (INHS).

Oecetis species D
NORTH CAROLINA: Columbus Co., L. Waccamaw @ L. Waccamaw, 1 male pupa, 3 females (ROM).

Oecetis species E
SOUTH CAROLINA: Barnwell Co., Castor Bay, Sav. R. Site, 1 male pupa, 2 females; Craig's Pond, Sav. R. Site, 1 male pupa (USNM), 2 males (INHS), 2 males (ROM).

Oecetis species F
FLORIDA: Osceola Co., L. Tohopekaliga @ Southport Pk., off Rt. 531, 1 female, 2 males (USNM).

Oecetis species G
ARIZONA: Pima Co., cow tank @ King Anvil Ranch, off Rt. 286, 1 female pupa, 1 male pupa (INHS), 1 female, 1 male (ROM).

LITERATURE CITED

Anderson, Norman H. 1974. Observations on the biology and laboratory rearing of *Pseudostenophylax edwardsi* (Trichoptera: Limnephilidae). Canadian Journal of Zoology 52: 7-13.

Banks, Nathan. 1895. New neuropteroid insects. Transactions of the American Entomological Society 22: 313-316.

Bjarnov, N. and J. Thorup. 1970. A simple method for rearing running-water insects, with some preliminary results. Archives Hydrobiologia. 67(2): 201-209.

Betten, Cornelius. 1934. The caddis flies or Trichoptera of New York state. New York State Museum Bulletin 292. 576 p.

Bilger, M. D. 1986. A preliminary checklist of the aquatic macroinvertebrates of New England, including New York state. U. S. Environ. Prot. Agency, Region I, Env. Services Div. 72 p.

Blickle, R. L. and W. J. Morse. 1966. The caddisflies (Trichoptera) of Maine excepting the family Hydroptilidae. Maine Agricultural Experiment Station Bulletin 24: 1-12.

Bowles, David E. and Michael L. Mathis. 1989. Caddisflies (Insecta: Trichoptera) of mountainous regions in Arkansas, with new state records for the order. Journal of the Kansas Entomological Society 2: 234-244.

_____. **1992.** A preliminary checklist of the caddisflies (Insecta: Trichoptera) of Oklahoma. Insecta Mundi 6: 29-35.

Chapin, Jay W. 1978. Systematics of Nearctic *Micrasema* (Trichoptera: Brachycentridae). Unpub. Ph.D. Dissertation. Clemson University. Clemson, South Carolina. 136 p.

Chen, Yianfang. 1993. Revision of the *Oecetis* (Trichoptera: Leptoceridae) of the world. Unpub. Ph.D. Dissertation. Clemson University. Clemson, South Carolina. 704 p.

Cobb, Donald G., J. F. Flannagan, and M. K. Friesen. 1984. Emergence of Trichoptera from two streams of the Duck Mountains in west-central Manitoba. Pages 75-87 *in* Proceedings of the Fourth International Symposium on Trichoptera Clemson, South Carolina. John C. Morse, editor. Series Entomologica 30, Dr. W. Junk Publishers, The Hague. 486 p.

Corbet, Phillip S., Fernand Schmid, and C. L. Augustin. 1966. The Trichoptera of St. Helen's Island, Montreal I. The species present and their relative abundance at light. Canadian Entomolologist 98: 1284-1298.

Cummins, Kenneth W. 1973. Trophic relations of aquatic insects. Annual Review of Entomology 18: 183-206.

Daigle, Gerald L. and James D. Haddock. 1981. The larval description and ecological notes of a caddisfly, *Nectopsyche tavara* (Ross) from the central Florida refugium (Trichoptera: Leptoceridae). Pan-Pacific Entomologist 57: 327-331.

Denning, Donald. G. 1951. Records and descriptions of Nearctic caddisflies, part III. Journal of the Kansas Entomological Society 24: 157-162.

Downs, Wilbur and P. A. Goodwin. 1987. Forest canopy collection of Trichoptera. Pages 207-209 *in* Proceedings of the Fifth International Symposium on Trichoptera, Lyon, France. M. Bournard and H. Tachet, editors. Series Entomologica 39, Dr. W. Junk Publishers, The Hague. 377 p.

Edwards, Sidney W. 1966. An annotated list of the Trichoptera of middle and west Tennessee. Tennessee Academy of Science 61: 116-128.

_____. **1973.** Texas caddisflies. The Texas Journal of Science 24: 491-516.

Ellis, Robert J. 1962. Adult caddisflies (Trichoptera) from Houghton Creek, Ogemaw County, Michigan. Occasional Papers of the Museum of Zoology, Univ. Michigan 624: 1-16.

Etnier, David A. 1965. An annotated list of the Trichoptera of Minnesota, with description of a new species. Entomological News 76: 141-152.

Etnier, David A. and Gunter A. Schuster. 1979. An annotated list of Trichoptera (caddisflies) of Tennessee. Journal of the Tennessee Academy of Science 54: 15-22.

Fischer, F. C. J. 1966. Trichopterorum Catalogus: 7. Nederlandse Entomologische Vereniging. Amsterdam, The Netherlands. 163 p.

_____. **1972.** Trichopterorum Catalogus: 14 (Suppl. Vols. 5, 6, & 7). Nederlandse Entomologische Vereniging. Amsterdam, The Netherlands. 154 p.

Flannagan, P. M. and J. F. Flannagan. 1982. Present distribution and the post-glacial origin of the Ephemeroptera, Plecoptera and Trichoptera of Manitoba. Manitoba Dept. of Natural Resources Technical Report 82-1, 79 p.

Flint, Oliver S. 1960. Taxonomy and biology of Nearctic limnephilid larvae (Trichoptera), with special reference to species in eastern United States. Entomologica Americana 40: 1-117.

_____. **1961.** The immature stages of the Arctopsychinae occurring in eastern North America Trichoptera: Hydropsychidae). Annals of the Entomological Society of America 54: 5-11.

_____. **1962.** Larvae of the caddis fly genus *Rhyacophila* in eastern North America (Trichoptera: Rhyacophilidae). Proceedings of the United States National Museum 113:465-493.

_____. **1964.** Notes on some Nearctic Psychomyiidae with special reference to their larvae (Trichoptera). Proceedings of the United States National Museum 115: 467-481.
_____. **1967.** Studies of Neotropical caddis flies, IV. New species from Mexico and Central America. Proceedings of the United States National Museum 123: 1-24.
_____. **1968.** Bredin-Archbold-Smithsonian biological survey of Dominica 9. the Trichoptera (caddisflies) of the Lesser Antilles. Proc. of the United States National Museum 125: 1-86.
_____. **1974.** The Trichoptera of Surinam, studies of neotropical caddisflies. XV. Stud. Foun. Surinam and the other Guyanas. No. 55. 14: 1-751.
_____. **1984.** The genus *Brachycentrus* in North America, with a proposed phylogeny of the genera of Brachycentridae (Trichoptera): Smithsonian Contrib. to Zoology 398: 1-58.
Flint, Oliver S. and W. L. Butler. 1983. *Hydropsyche brunneipennis*, new species, a member of the *scalaris* Group, from the Potomac River near Washington, D.C. (Trichoptera: Hydropsychidae). Proc. of the Entomological Society of Washington 85: 205-211.
Floyd, Michael A., John C. Morse, and James V. McArthur. 1993. Aquatic insects of Upper Three Runs Creek, Savannah River Site, South Carolina. Part IV: caddisflies (Trichoptera) of the lower reaches. Journal of Entomological Science 28: 85-95.
Floyd, Michael A. and G. A. Schuster. 1990. The caddisflies (Insecta: Trichoptera) of the Buck Creek system, Pulaski County, Kentucky. Transactions of the Kentucky Academy of Science 51: 127-134.
Frazer, Kenneth S., Steven C. Harris, and G. Milton Ward. 1991. Survey of the Trichoptera in the Little River drainage of northeastern Alabama. Bull. Alabama Mus. Nat. Hist. 11: 17-22.
Garono, Ralph J. and David B. MacLean. 1988. Caddisflies (Trichoptera) of Ohio wetlands as indicated by light-trapping. Ohio Journal of Science 88: 143-151.
Glover, James B. 1993. The taxonomy and biology of the larvae of the North American caddisflies in the genera *Triaenodes* and *Ylodes* (Trichoptera: Leptoceridae). Unpub. Ph.D. Dissertation. University of Louisville, Louisville, Kentucky. 252 p.
Glover, James B. and Donald C. Tarter. 1990. The Leptoceridae (Trichoptera) of West Virginia. Entomological News 101: 35-38.
Gordon, Elizabeth A. 1984. The Trichoptera of Florida: a preliminary survey. Pages 161-166 *in* Proceedings of the Fourth International Symposium on Trichoptera, Clemson, South Carolina. J. C.Morse, editor. Series Entomologica 30, Dr. W. Junk Publishers, The Hague. 486 p.
Haag, Kathy H., Vincent H. Resh, and Stuart E. Neff. 1984. Changes in the adult caddisfly (Trichoptera) community of the Salt River, Kentucky. Transactions of the Kentucky Academy of Science 45: 101-108.
Haddock, James D. 1977. The biosystematics of the caddisfly genus *Nectopsyche* in North America with emphasis on the aquatic stages. American Midland Naturalist 98: 382-421.
Hamilton, Steven W., Gunter A. Schuster, and M. B. DuBois. 1983. Checklist of the Trichoptera of Kansas. Transactions of the Kansas Academy of Science 86: 10-23.
Harris, Todd L. 1971. Crepuscular flight periodicity of Trichoptera. Journal of the Kansas Entomological Society 44: 293-301.
Harris, Steven C. 1990. Preliminary consideration on rare and endangered invertebrates in Alabama. Journal of the Alabama Academy of Science 61: 64-92.
Harris, Steven C. and Robert W. Kelley. 1983. New Hydroptilidae (Trichoptera) from Alabama and South Carolina. Entomological News 94: 181-186.
Harris, Steven C. and Paul K. Lago. 1990. Annotated checklist of the Rhyacophiloidea and Integripalpia (Trichoptera) of Alabama. Entomological News 101: 57-66.
Harris, Steven C., Paul K. Lago, and R. B. Carlson. 1980. Preliminary survey of the Trichoptera of North Dakota. Proc. of the Entomological Society of Washington 82: 39-43.
Harris, Steven C., Paul K. Lago, and Patrick E. O'Neil. 1984. Trichoptera of the Cahaba River system in Alabama. Entomological News 95: 103-112.
Harris, Steven C., Paul K. Lago, and Joseph F. Scheiring. 1982. An annotated list of Trichoptera of several streams on Eglin Air Force Base, Florida. Entomological News 93: 79-84.
Harris, Steven C., Patrick E. O'Neil, R. V. Chandler, M. F. Mettee, E. J. McCullough. 1983. Biological and hydrological impacts of surface mining for federal minerals on the Tyro Creek watershed, Alabama. Phase I. Premining-aquatic baseline information. Geological Survey of Alabama, 98 p.
Harris, Steven C., Patrick E. O'Neil, and Paul K. Lago. 1991. Caddisflies of Alabama. Geological Survey of Alabama Bulletin 142: 1-442.
Hennig, Willi. 1966. Phylogenetic Systematics, translated by D. Davis and R. Zangerl. University of Illinois Press. Urbana, Illinois. 263 p.
Herrman, Scott J., Dave E. Ruiter, and John D. Unzicker. 1986. Distribution and records of Colorado Trichoptera. The Southwestern Naturalist 31: 421-457.
Hiley, P. O. 1969. A method of rearing Trichoptera larvae for taxonomic purposes. Entomological Monograph Magazine 105: 278-279.
Hilsenhoff, William L. 1975. Aquatic insects of Wisconsin. Keys to Wisconsin genera and notes on bi-

ology, distribution, and species. Publication of the Natural History Council, University of Wisconsin. Madison, Wisconsin. 60 p.

Hilsenhoff, William L., J. L. Longridge, Richard P. Narf, Kenneth J. Tennessen, and C. P. Walton. 1972. Aquatic insects of the Pine-Pople River, Wisconsin. Wisconsin Department of Natural Resources Bulletin 54: 1-44.

Holzenthal, Ralph W. 1982. The caddisfly genus *Setodes* in North America (Trichoptera: Leptoceridae). Journal of the Kansas Entomological Society 55: 253-271.

Holzenthal, Ralph W., Steven C. Harris, and Paul K. Lago. 1982. An annotated checklist of the caddisflies (Trichoptera) of Mississipi and southeastern Louisiana. Part III: Limnephiloidea and conclusions. Proc. of the Entomological Society of Washington 84: 513-520.

Holzenthal, Ralph W. and Robert W. Kelley. 1983. New micro-caddisflies from the southeastern United States (Trichoptera: Hydroptilidae). Florida Entomologist 66: 464-472.

Huryn, Alex D. and Benjamin A. Foote. 1983. An annotated list of the caddisflies (Trichoptera) of Ohio. Proceedings of the Entomological Society of Washington 85: 783-796.

Jones, T. S. and Vincent H. Resh. 1988. Movements of adult aquatic insects along a Montana (USA) springbrook. Aquatic Insects 10: 99-104.

Kelley, Robert W. and John C. Morse. 1982. A key to the females of the genus *Oxyethira* (Trichoptera: Hydroptilidae) from the southern United States. Proc. of the Entomological Society of Washington 84: 256-269.

Lager, T. M., M. D. Johnson, S. N. Williams, and J. L. McCulloch. 1979. A preliminary report on the Plecoptera and Trichoptera of northeastern Minnesota. The Great Lakes Entomologist 12: 109-114.

Lake, R. W. 1984. Distribution of caddisflies (Trichoptera) in Delaware. Entomological News 95: 215-224.

Lenat, David R. 1988. Water quality assessment of streams using a qualitative collection method for benthic macroinvertebrates. Journal of the North American Benthological Society 7: 222-233.

Lenat, David R. and David L. Penrose. 1980. Discussion of "Hierarchical diversity of communities of aquatic insects and fishes" by R. L. Kaesier and E. E. Herricks. Water Resources Bulletin 16: 361-362.

Lepneva, S. G. 1966. Fauna of the U.S.S.R.; Trichoptera, vol. 2, no. 2. Larvae and Pupae of Integripalpia. Published by the Israei Program for Scientific Translations, 1970.

Longridge, J. L. and William L. Hilsenhoff. 1973. Annotated list of Trichoptera (caddisflies) in Wisconsin. Wisconsin Academy of Science, Arts & Letters 61: 173-183.

MacLachlan, Robert. 1877. A monographic revision and synopsis of the Trichoptera of the European fauna. pt. 6: 281-348.

Masteller, Edward C. and Oliver S. Flint. 1979. Light trap and emergence trap records of caddis flies (Trichoptera) of the Lake Erie region of Pennsylvania and adjacent Ohio. The Great Lakes Entomologist 12: 165-177.

_____. **1992.** The Trichoptera (caddisflies) of Pennsylvania: an annotated checklist. Journal of the Pennsylvania Academy of Science 66: 68-78.

Mathis, Michael L. and David E. Bowles. 1992. A preliminary survey of the Trichoptera of the Ozark Mountains, Missouri, U.S.A. Entomological News 103: 19-29.

McCabe, T. L. 1986. An annotated list of pine bush caddis. Skenectada 3:17-18.

McElravy, Eric P., Thomas L. Arsuffi, and Benjamin A. Foote. 1977. New records of caddisflies (Trichoptera) for Ohio. Proceedings of the Entomological Society of Washington 79: 599-604.

Mecom, J. O. and Kenneth W. Cummins. 1964. A preliminary study of the trophic relationships of the larvae of *Brachycentrus americanus* (Banks) (Trichoptera: Brachycentridae). Transactions of the American Microscopic Society 83: 233-43.

Milne, Lorus J. 1934. Studies in North American Trichoptera 1: 1-128.

Milne, Margaret J. 1938. The "metamorphotype method" in Trichoptera. Journal of the New York Entomological Society 46: 435-437.

Mingo, T. M. and Elizabeth K. Gibbs. 1980. The aquatic insects of the Narraguagus River, Hancock and Washington Counties, Maine. Life Sciences Agricultural Experiment Station, Univ. Maine, Orono, Bulletin 100: 1-63.

Morse, John C. 1975. A phylogeny and revision of the caddisfly genus *Ceraclea* (Trichoptera: Leptoceridae). Contributions of the American Entomological Institute 11: 1-97.

_____. **1982.** Trichoptera. Pages 51-71 *in* Rare and Endangered Biota of Florida: Volume 6, Invertebrates. Richard Franz, editor. Univ. Presses of Florida. Gainesville, Florida.

_____. **1993.** A checklist of the Trichoptera of North America, including Greenland and Mexico. Transactions of the American Entomological Society 119: 47-93.

Morse, John C., Jay W. Chapin, David D. Herlong, and R. S. Harvey. 1980. Aquatic insects of Upper Three Runs Creek, Savannah River Plant, South Carolina. Part 1: Orders other than Diptera. Journal of the Georgia Entomological Society 15: 73-101.

Morse, John C., Steven W. Hamilton, and K. M. Hoffman. 1989. Aquatic insects of Lake Jocassee Catchment in North and South Carolina, with descriptions of four new species of caddisflies

(Trichoptera). The Journal of the Elisha Mitchell Society 105: 14-33.

Morse, John C. and Ian D. Wallace. 1976. *Athripsodes* Billberg and *Ceraclea* Stephens, distinct genera of long-horned caddisflies. Pages 33-40 *in* Proc. of the First International Symposium on Trichoptera, Lunz, Austria. H. Malicky, editor. Dr. W. Junk Publishers. The Hague. 213 p.

Moulton, Stephen R., II and Kenneth W. Stewart. 1993. Nearctic *Oecetis scala* Group (Trichoptera: Leptoceridae) with the description of two new species from the Interior Highlands. Annals of the Entomological Society of America 86: 221-227.

Moulton, Stephen R., II, D. Petr, and Kenneth W. Stewart. 1993. Caddisflies (Insecta: Trichoptera) of the Brazos River drainage in north-central Texas. The Southwestern Naturalist 38: 19-23.

Moulton, Stephen R., Kenneth W. Stewart, and K. L. Young. 1994. New records, distribution and taxonomic status of some northern Arizona caddisflies (Trichoptera). Entomological News 105: 164-174.

Nelson, V. A. 1987. Light trap collections of Trichoptera near the source of Hop Brook, New Haven County, Connecticut. Pages 203-205 *in* Proceedings of the Fifth International Symposium on Trichoptera, Lyon, France. M. Bournard and H. Tachet, editors. Series Entomologica 39. Dr. W. Junk Publishers. The Hague. 377 p.

Neves, Richard J. 1979. A checklist of caddisflies (Trichoptera) from Massachusetts. Entomological News 90: 167-175.

Newell, Robert L. and Dave S. Potter. 1973. Distribution of some Montana caddisflies (Trichoptera). Proceedings of the Montana Academy of Science 33: 12-21.

Nimmo, Andrew P. 1966. A list of Trichoptera taken at Montreal and Chambly, Quebec, with descriptions of three new species. Canadian Entomologist 98: 688-693.

Nimmo, Andrew P. and G. G. E. Scudder. 1978. An annotated checklist of the Trichoptera (Insecta) of British Columbia. Syesis 11: 117-133.

_____. **1983.** Supplement to an annotated checklist of the Trichoptera (Insecta) of British Columbia. Syesis 16: 71-83.

Parker, Charles R. and J. Reese Voshell, Jr. 1981. A preliminary checklist of the caddisflies (Trichoptera) of Virginia. Journal of the Georgia Entomological Society 16: 1-7.

Peterson, C. and Benjamin A. Foote. 1980. Annotated list of Trichoptera collected along Furnace Run of the Cuyahoga Valley National Recreation Area in northeastern Ohio. The Great Lakes Entomologist 13: 201-205.

Philipson, G. N. 1953. A method of rearing trichopterous larvae collected from swiftflowing waters. Proc. of the Royal Entomological Society of London (A), 28: 15-16.

Resh, Vincent H. 1972. A technique for rearing caddisflies (Trichoptera). Canadian Entomologist 104: 1959-1961.

_____. **1975.** A distributional study of the caddisflies of Kentucky. Transactions of the Kentucky Academy of Science 36: 6-16.

_____. **1976.** The biology and immature stages of the caddisfly genus *Ceraclea* in eastern North America (Trichoptera: Leptoceridae). Annals of the Entomological Society of America 69: 1039-1061.

Resh, Vincent H., Kathy H. Haag, and Stuart E. Neff. 1975. Community structure and diversity of caddisfly adults from the Salt River, Kentucky. Environmental Entomology 4: 241-253.

Resh, Vincent H., David M. Rosenberg, and A. P. Wiens. 1983. Emergence of caddisflies (Trichoptera) from eroding and non-eroding shorelines of Southern Indian Lake, Manitoba, Canada. Canadian Entomologist 115: 1563-1572.

Resh, Vincent H. and John D. Unzicker. 1975. Water quality monitoring and aquatic organisms: the importance of species identification. Journal of Water Pollution Control Federation 47: 9-19.

Resh, Vincent H., David S. White, and S. J. White. 1978. Lake Texoma caddisflies (Insecta: Trichoptera): 1. Species present and faunal changes since impoundment. SouthwesternNaturalist. 23: 381-388.

Richardson, John S. and Hugh F. Clifford. 1986. Phenology and ecology of some Trichoptera in a low-gradient boreal stream. Journal of the North American Benthological Society 5: 191-199.

Roemhild, G. 1982. The Trichoptera of Montana with distributional and ecological notes. Northwest Science 56: 8-13.

Ross, Herbert H. 1941. Descriptions and records of North America Trichoptera. Transactions of the American Entomological Society 67: 35-126.

_____. **1944.** The caddisflies, or Trichoptera, of Illinois. Bulletin of the Illinois Natural History Survey 23: 1-236.

_____. **1966.** Two new species of *Oecetis* occurring in eastern North America (Trichoptera: Leptoceridae). Transactions Illinois State Academy of Science 59: 11-14.

Ross, Herbert H. and J. Bruce Wallace. 1974. The North American genera of the family Sericostomatidae (Trichoptera). Journal of the Georgia Entomological Society 9: 42-48.

Roy, D. and Peter P. Harper. 1975. Nouvelles mentions de trichoptères du Québec et description de *Limnephilus nimmoi* sp. nov. (Limnephilidae). 53: 1080-1088.

Roy, D. and Peter P. Harper. 1979. Liste préliminaire des Trichoptères (insectes) du Québec. Annales de la Société Entomologique du Québec 24: 148-172.

Ruiter, Dave E. and Robert J. Lavigne. 1985. Distribution of Wyoming Trichoptera. Agricultural Experiment Station, University of Wyoming. 102p.

Schefter, Patricia W. and Glenn B. Wiggins. 1986. A systematic study of the Nearctic larvae of the *Hydropsyche morosa* group (Trichoptera: Hydropsychidae). Life Sciences Miscellaneous Publications of the Royal Ontario Museum. Toronto, Canada. 94 p.

Schmid, Fernand. 1979. On some new trends in Trichopterology. Bulletin of the Entomological Society of Canada 11: 48-57.

_____. **1980.** Les insectes et arachnides du Canada. Partie 7. Genera des Trichoptera du Canada et des Estes adjacents. Pub. 1692, Agric. Canada. 296 p.

_____. **1984.** Essai d'evaluation de la faune mondiale des trichoptères. Page 337 *in* Proceedings of the Fourth International Symposium on Trichoptera, Clemson, South Carolina. John C. Morse, editor. Series Entomologica 30. Dr. W. Junk Publishers. The Hague. 486 p.

Schuster, Gunter A. and David A. Etnier. 1978. A manual for the identification of the larvae of caddisfly genera *Hydropsyche* Pictet and *Symphitopsyche* Ulmer in eastern and central North America (Trichoptera: Hydropsychidae). U. S. Environ. Prot. Agency, Environ. and Suport Lab., EPA-600/4-78-060, 129 p.

Shapas, T. J. and William L. Hilsenhoff. 1976. Feeding habits of Wisconsin's predominant lotic Plecoptera, Ephemeroptera, and Trichoptera. Great Lakes Entomologist 9: 175-188.

Sherberger, Fred F. and J. B. Wallace. 1971. Larvae of the southeastern species of *Molanna*. Journal of the Kansas Entomological Society 44: 217-224.

Smith, D. H. and D. M. Lehmkuhl. 1980. Analysis of two problematic North American caddisfly species: *Oecetis avara* (Banks) and *Oecetis disjuncta* (Banks) (Trichoptera: Leptoceridae). Quaestiones Entomologicae 16: 641-656.

Sparks, A. N., R. D. Jackson, James E. Carpenter, and R. A. Muller. 1986. Insects captured in light traps in the Gulf of Mexico. Annals of the Entomological Society America 79: 132-139.

Steven, Jeffrey C. and William L. Hilsenhoff. 1984. The caddisflies (Trichoptera) of Otter Creek, Wisconsin. Wisconsin Academy of Science, Arts and Letters 72: 157-172.

Swegman, Bernard G., W. Walker, and Jan L. Sykora. 1981. The adult Trichoptera of Linesville Creek, Crawford County, Pennsylvania with notes on their flight activity. Transactions of the American Entomological Society 107: 125-147.

Tarter, Donald C. 1990. A checklist of the caddisflies (Trichoptera) from West Virginia. Entomological News 101: 236-245.

Unzicker, John D., L. Aggus, and L. O. Warren. 1970. A preliminary list of the Arkansas Trichoptera. Journal of the Georgia Entomological Society 5: 167-174.

Unzicker, John D., Vincent H. Resh, and John C. Morse. 1982. Trichoptera. Pages 9.1-9.138 *in* Aquatic insects and oligochaetes of North and South Carolina. A. R. Brigham, W. U. Brigham, and A. Gnilka, editors. Midwest Aquatic Enterprises. Mahomet, Illinois. 837 p.

Usis, John D. and David B. MacLean. 1986. The caddisflies (Trichoptera) of Stillfork Swamp Nature Preserve, Carroll County, Ohio. The Ohio Journal of Science 86: 33-40.

Vorhies, C. T. 1909. Studies of the Trichoptera of Wisconsin. Wisconsin Academy of Science Transactions 16: 647-738.

Wallace, Ian D. 1981. A key to larvae of the family Leptoceridae (Trichoptera) in Great Britain and Ireland. Freshwater Biology 11: 273-297.

Wallace, Ian D., B. Wallace, and G. N. Philipson. 1990. A key to the case-bearing caddis larvae of Britain and Ireland. Freshwater Biology Association of Ambleside. Cumbria, England XXX p.

Wallace, J. Bruce. 1971. A new species of *Brachycentrus* from Georgia with two unusual larval characters (Trichoptera: Brachycentriidae). Entomological News 82: 313-321.

Wallace, J. Bruce and F. F. Sherberger. 1970. The immature stages of *Anisocentropus pyraloides* (Trichoptera: Calamoceratidae). Journal of the Georgia Entomological Society 5: 217-224.

Waltz, Robert D. and W. Patrick McCafferty. 1983a. The caddisflies of Indiana. Purdue Univ. Agricultural Experiment Station Bulletin 978: 1-25.

_____. **1983b.** New caddisfly records for New Mexico (Insecta: Trichoptera). The Southwestern Naturalist 28: 353-356.

Wiggins, Glenn B. 1959. A method of rearing caddisflies (Trichoptera). Canadian Entomologist 91: 402-405.

_____. **1960.** A preliminary systematic study of the North American larvae of the caddisfly family Phryganeidae (Trichoptera). Canadian Journal of Zoology 38: 1153-1170.

_____. **1966.** The critical problem of systematics in stream ecology. Pages 52-58 *in* K. W. Cummins, C. A. Fryon, Jr., and R. T. Hartman,editors, Symposium on organism-substrate relationships in streams. Special Publication 4. Pymatuning Laboratory of Field Ecology. Pittsburgh, Pennsylvania.

_____. **1973a.** New systematic data for the North American caddisfly genera *Lepania*, *Goeracea*, and *Goerita* (Trichoptera: Limnephilidae). Life Science Contributions, Royal Ontario Museum 91.

_____. **1973b.** Contributions to the systematics of the caddisfly family Limnephilidae (Trichoptera). I.

LifeScience Contributions, Royal Ontario Museum 94.

_____. **1977.** Larvae of the North American caddisfly genera (Trichoptera). University of Toronto Press, Toronto, Canada. 401 p.

_____. **1981.** Considerations on the relevance of immature stages to the systematics of Trichoptera. Pages 395-407 *in* Proceedings of the Third International Symposium on Trichoptera, Perugia, Italy. G. P. Morettii, editor. Series Entomologica 20. Dr. W. Junk Publishers. The Hague. 472 p.

Wiggins, Glenn B. and Norman H. Anderson. 1968. Contributions to the systematics of the caddisfly genera *Pseudostenophylax* and*Philocasca* with special reference to the immature stages (Trichoptera: Limnephilidae). Canadian Journal of Zoology 46: 61-75.

Wimmer, H. P. 1979. An anatomical and phylogenetic study of Trichoptera, includinga partial check-list of Vermont species. Unpub. Ph.D. Dissertation. University of Vermont. Burlington, Vermont,, 300 p.

Winterbourn, Michael J. 1971. The life histories and trophic relationships of the Trichoptera of Marion Lake, British Columbia. Canadian Journal of Zoology 49: 623-635.

Yammamoto, T. and Glenn B. Wiggins. 1964. A comparative study of the North American species in the caddisfly genus *Mystacides* (Trichoptera: Leptoceridae). Canadian Journal of Zoology 42: 1105-1126.

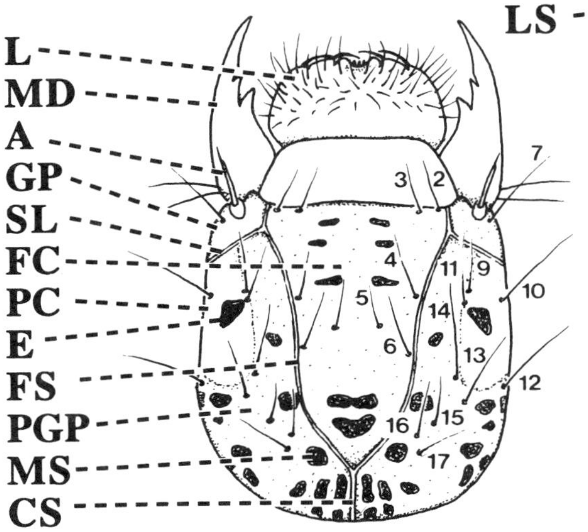

1

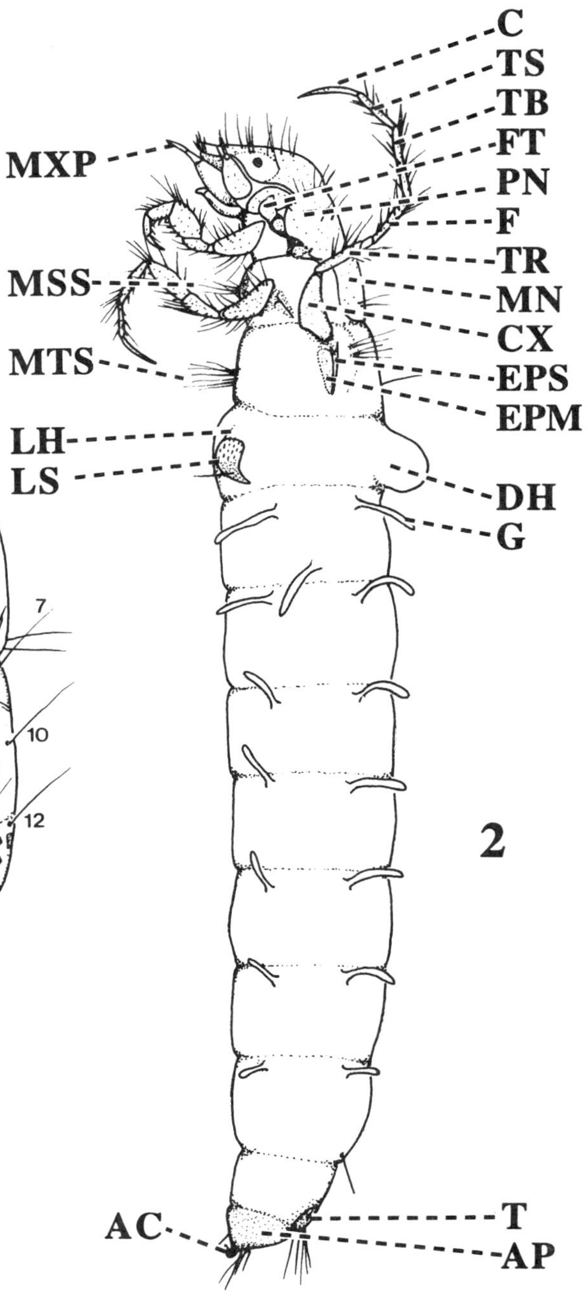

2

Figure 1. Generalized larval head of *Oecetis*, dorsal view. Abbreviations: A, antenna; CS, coronal suture; E, eye; FC, frontoclypeus; FS, frontoclypeal suture; GP, genal parietal sclerite; L, labrum; MD, mandible; MS, muscle scar; PC, periocular corona; PGP, postgenal parietal sclerite; SL, subocular line; numerals refer to head setae using the designations of Wiggins (1977).

Figure 2. Generalized *Oecetis* larva, left lateral view. Abbreviations: AC, anal claw; AP, anal proleg; C, tarsal claw; CX, coxa; DH, dorsal hump of abdominal segment I; EPM, epimeron; EPS, episternum; F, femur; FT, foretrochantin; G, gill; LH, lateral hump of abdominal segment I; LS, sclerite of lateral hump; MN, mesonotum; MSS, mesosternal setae; MXP, maxillary palp; MTS, metasternal setae; PN, pronotum; T, tergite of abdominal segment IX; TB, tibia; TR, trochanter; TS, tarsus.

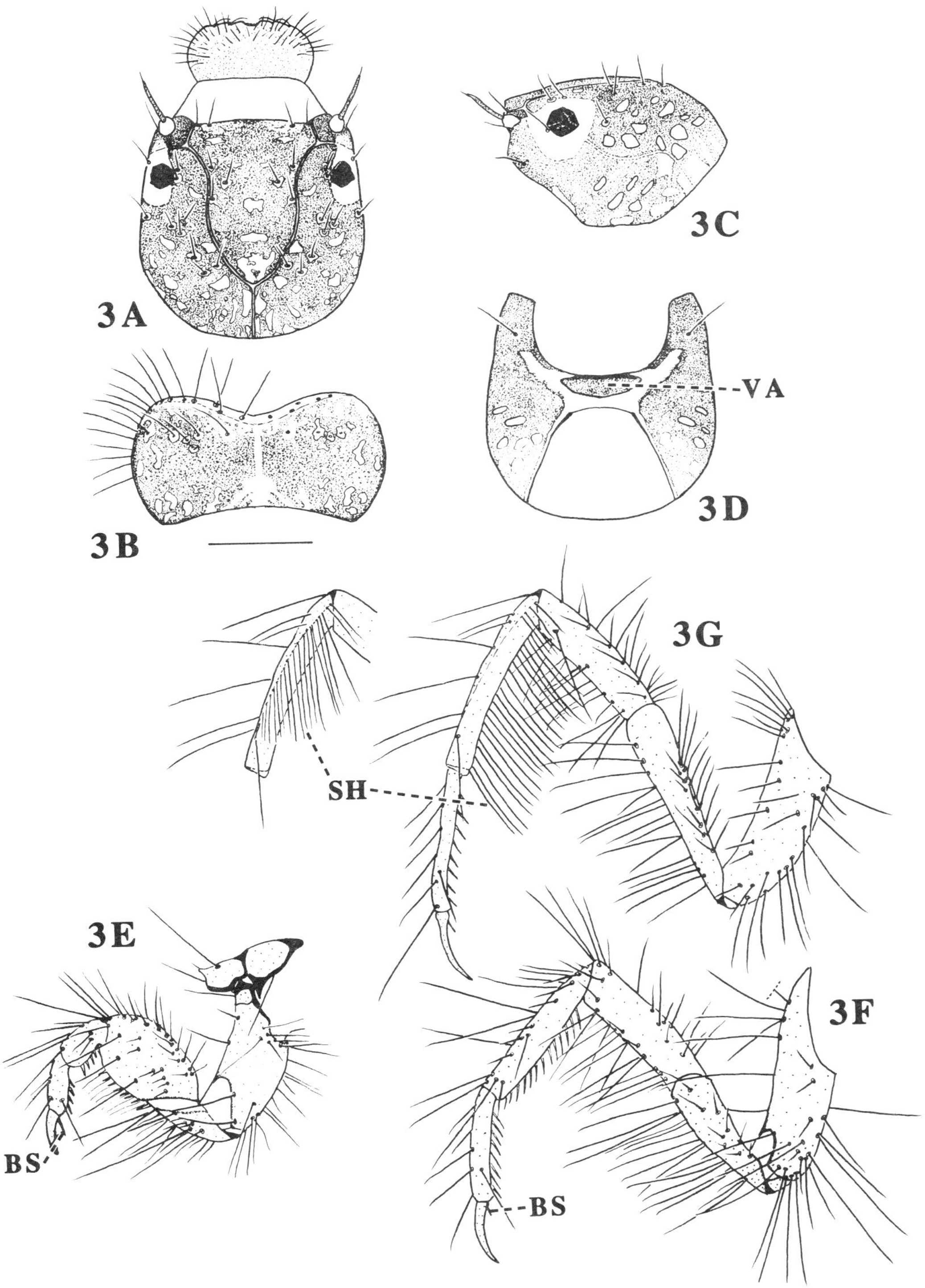

Figure 3. *Oecetis cinerascens*, larva and case. (A) head, dorsal; (B) pronotum, dorsal (scale bar = 0.5 mm); (C) head, left lateral; (D) head, ventral, showing ventral apotome, VA; (E) left foreleg, lateral, showing basal seta, BS; (F) left midleg, lateral, showing basal seta, BS; (G) left hindleg, showing swimming hairs, SH.

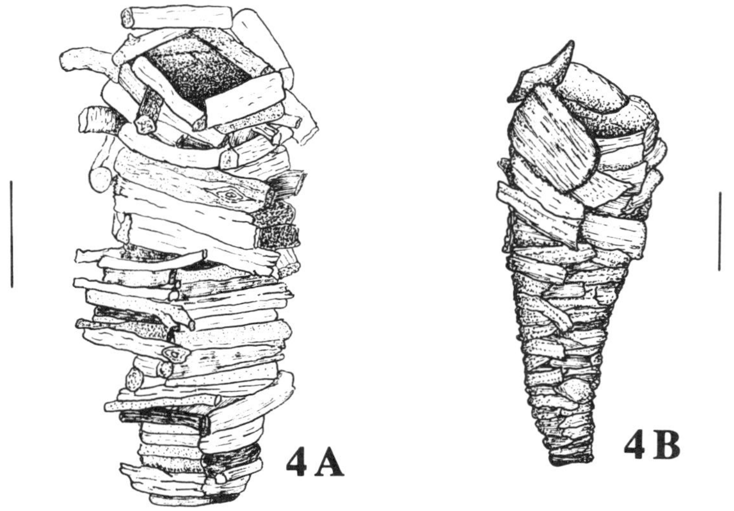

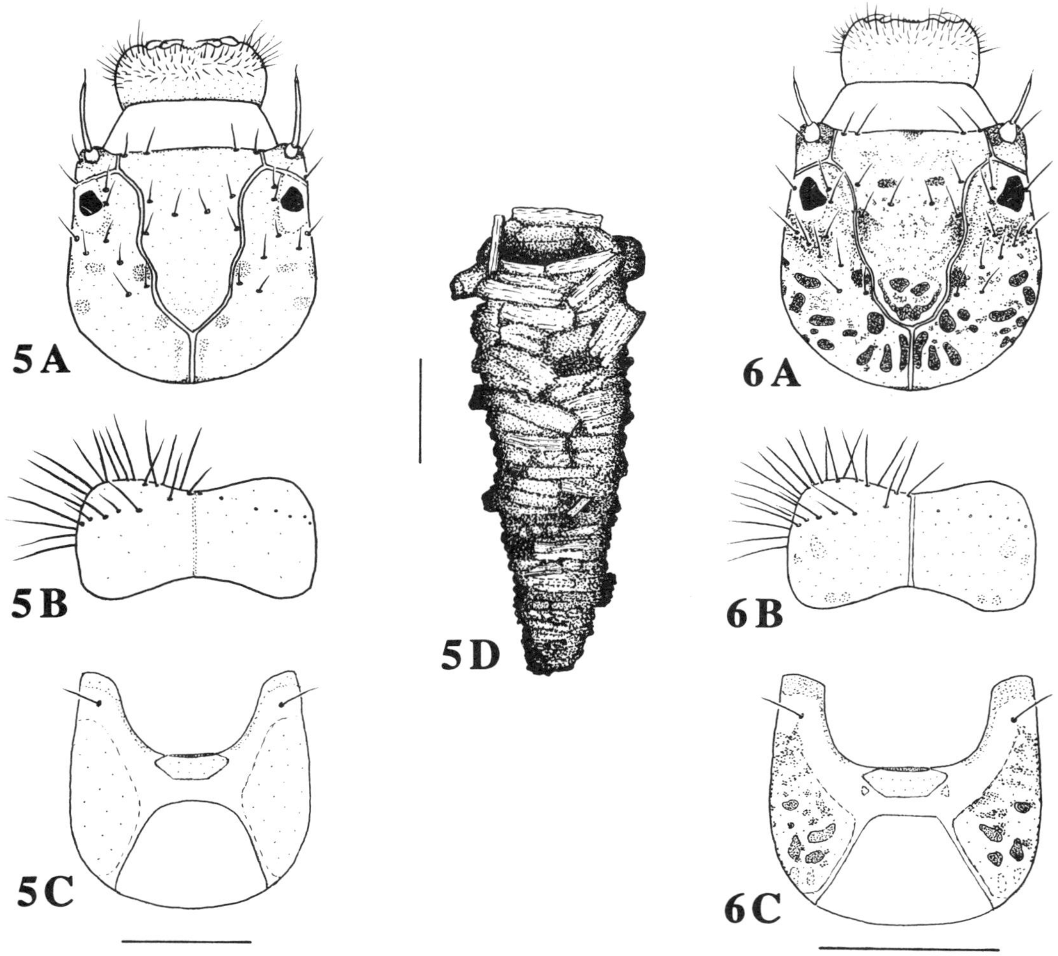

Figure 4. *Oecetis cinerascens*. (A-B) cases (scale bars = 1.5 mm).

Figure 5. *Oecetis georgia*, larva and case. (A) head, dorsal; (B) pronotum, dorsal; (C) head, ventral (scale bar = 0.5 mm); (D) case (scale bar = 1.5 mm).

Figure 6. *Oecetis persimilis*, larva. (A) head, dorsal; (B) pronotum, dorsal; (C) head, ventral (scale bar = 0.5 mm).

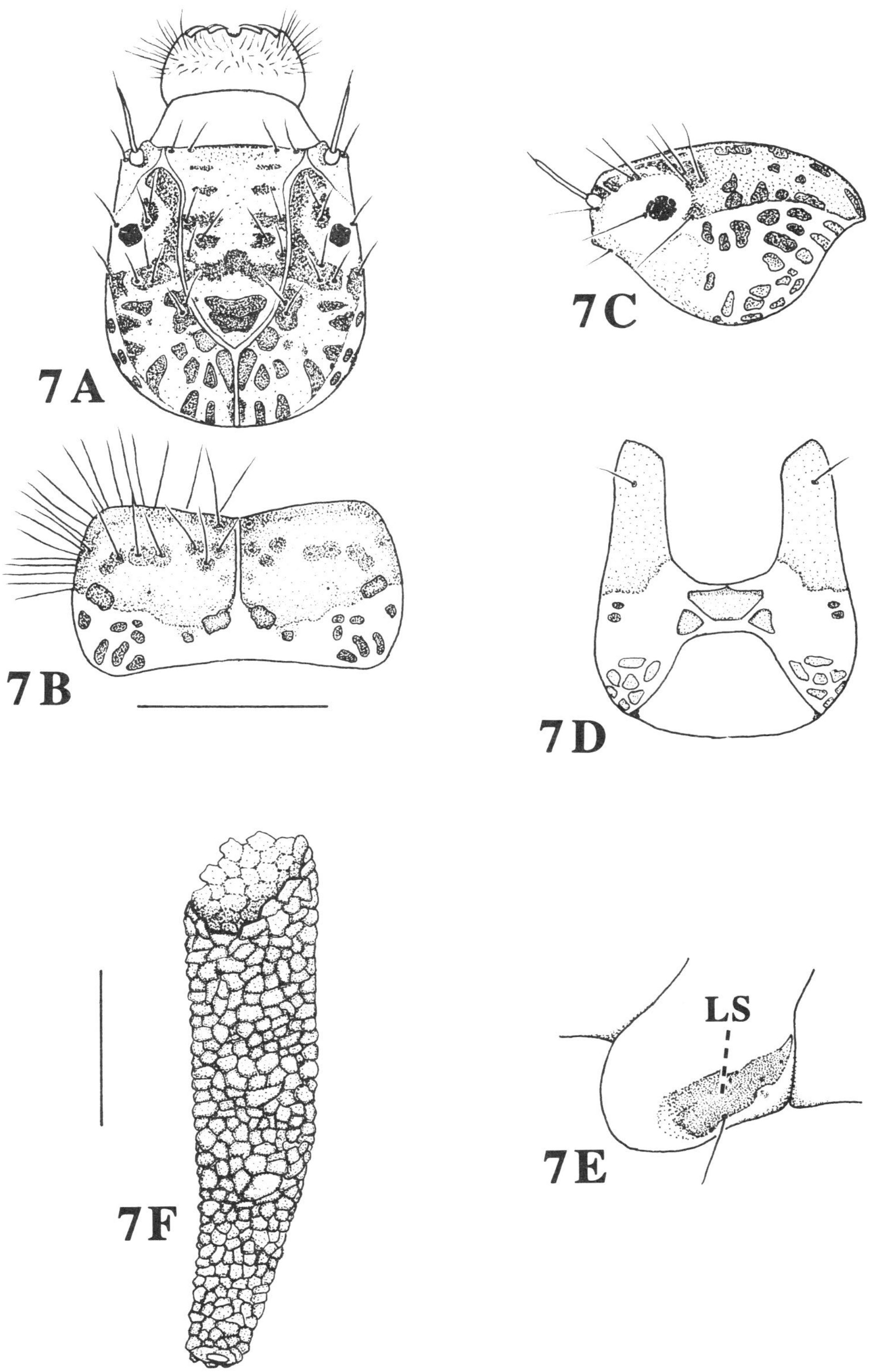

Figure 7. *Oecetis morsei*, larva and case. (A) head, dorsal; (B) pronotum, dorsal (scale bar = 0.5 mm); (C) head, left lateral; (D) head, ventral; (E) left lateral hump of abdominal segment I, showing lateral sclerite, LS; (F) case (scale bar = 1.5 mm).

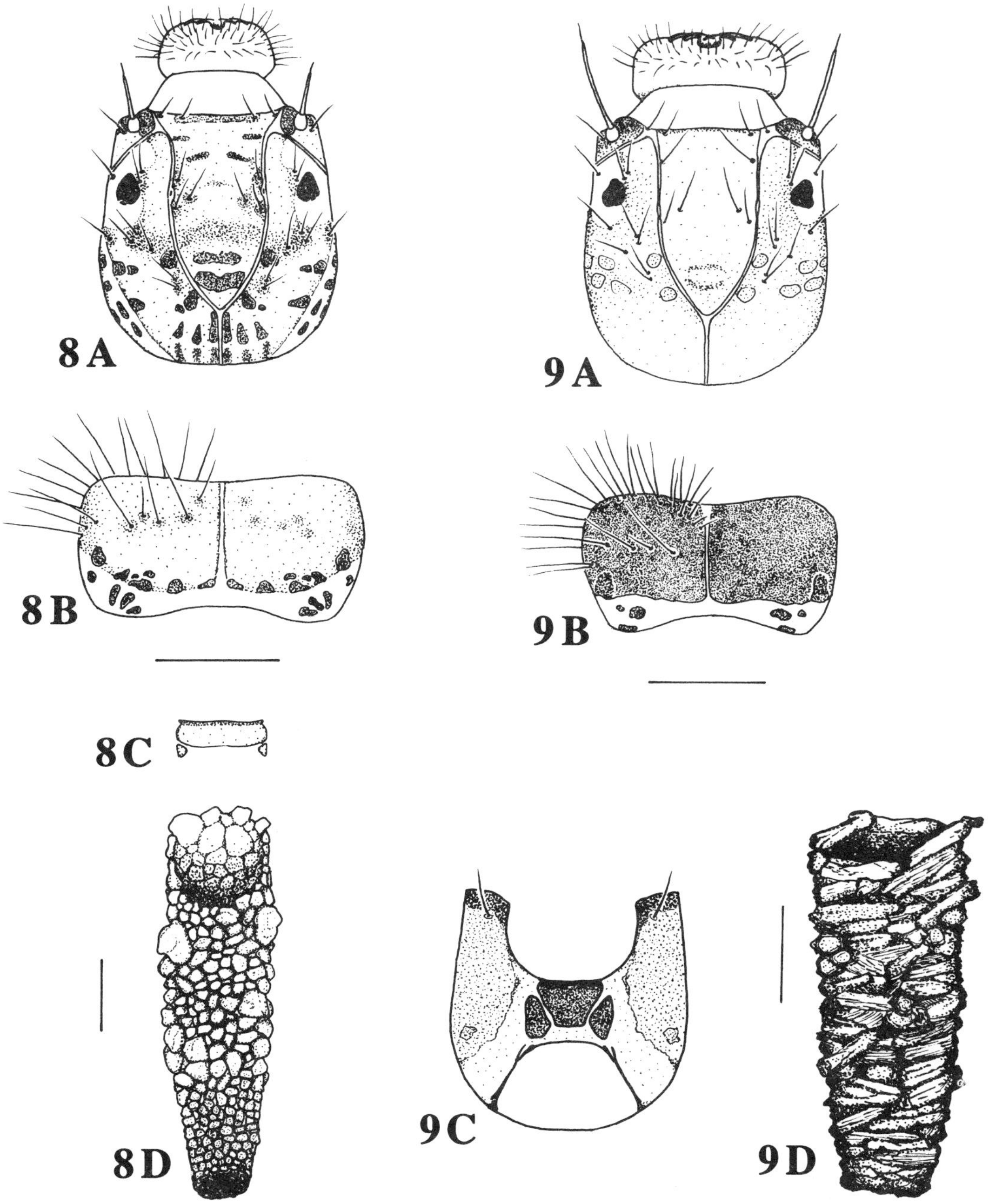

Figure 8. *Oecetis eddlestoni*, larva and case. (A) head, dorsal; (B) pronotum, dorsal (scale bar = 0.5 mm); (C) ventral apotome; (D) case (scale bar = 1.5 mm).

Figure 9. *Oecetis arizonica*, larva and case. (A) larval head, dorsal; (B) pronotum, dorsal (scale bar = 0.5 mm); (C) larval head, ventral; (D) case (scale bar = 1.5 mm).

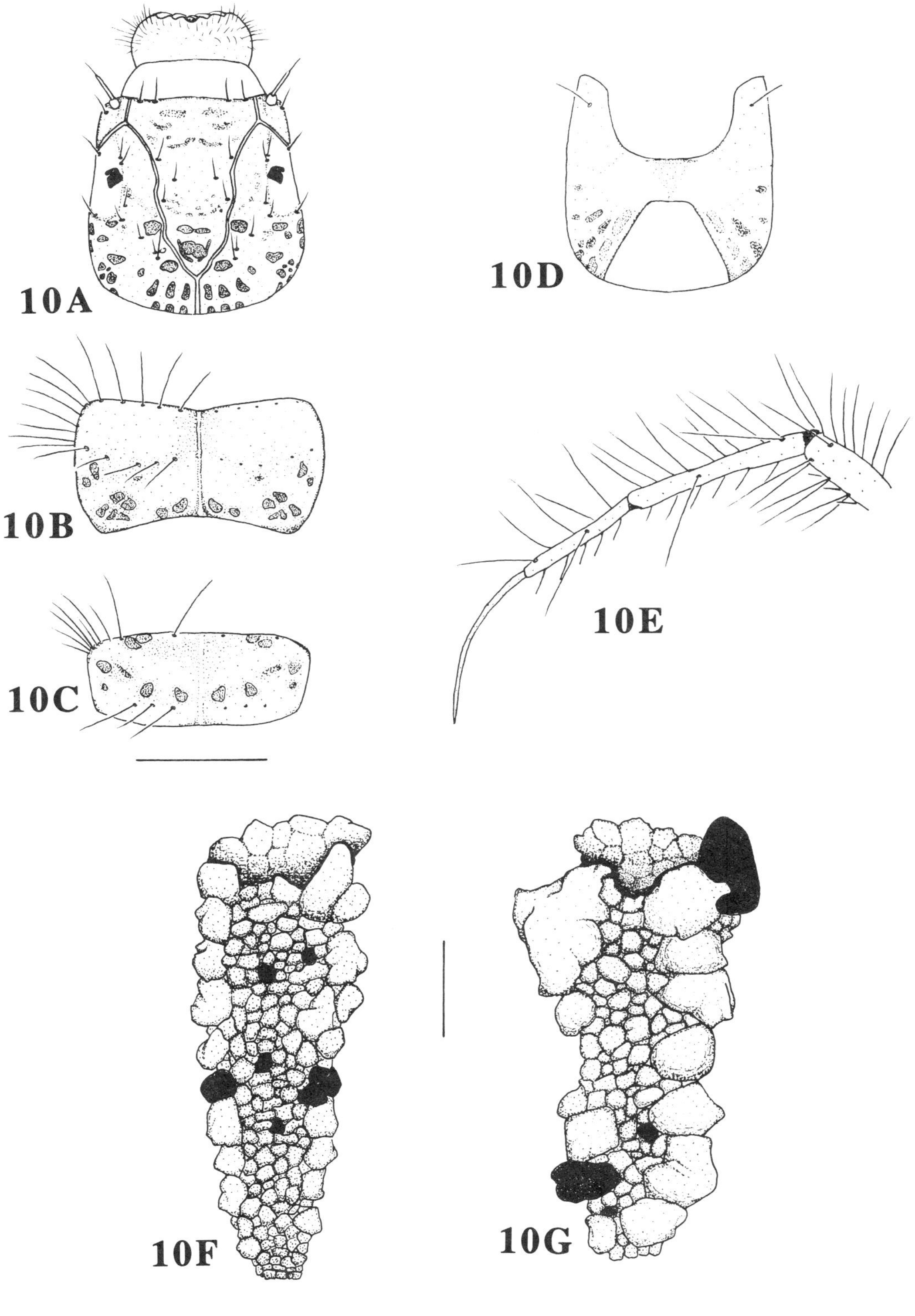

Figure 10. *Oecetis nocturna*, larva and cases. (A) head, dorsal; (B) pronotum, dorsal; (C) mesonotum, dorsal (scale bar = 0.5 mm); (D) head, ventral; (E) left hindleg, lateral; (F-G) cases (scale bar = 1.5 mm).

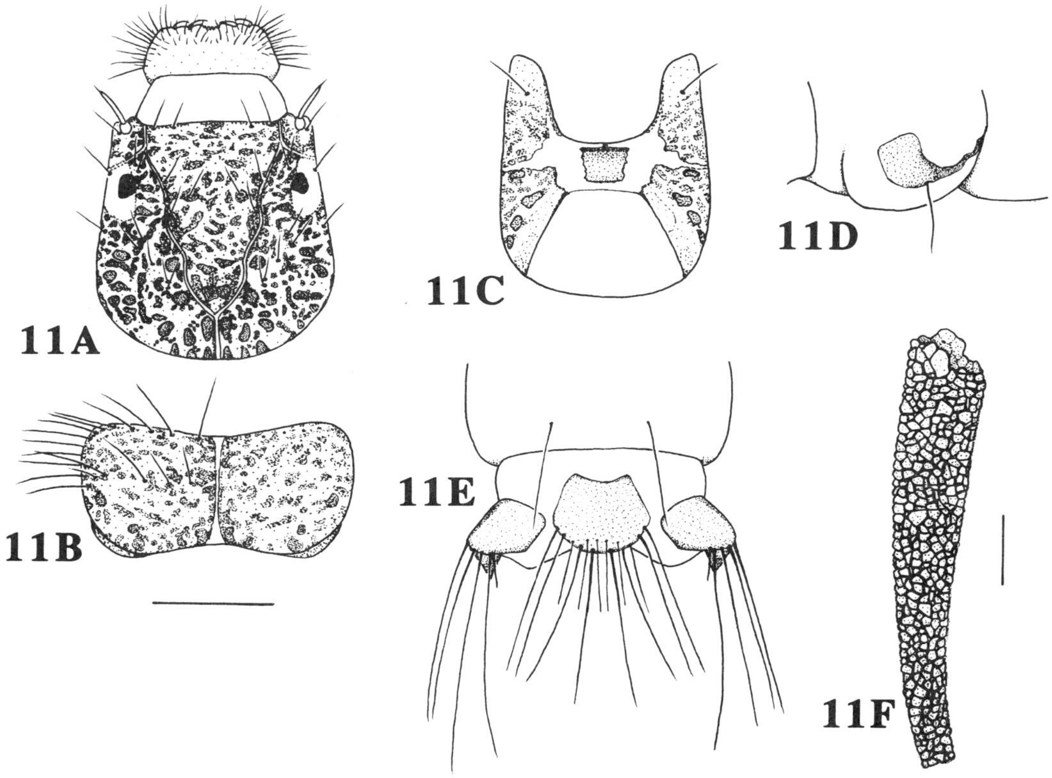

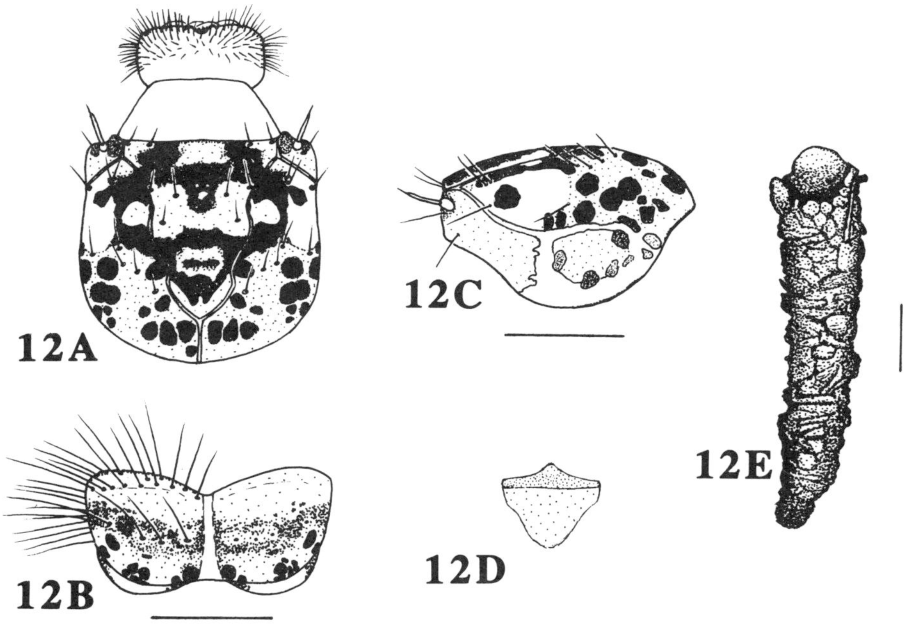

Figure 11. *Oecetis porteri*, larva and case. (A) head, dorsal; (B) pronotum, dorsal (scale bar = 0.5 mm); (C) head, ventral; (D) left lateral hump, showing lateral sclerite, LS; (E) tergite of abdominal segment IX, dorsal; (F) case (scale bar = 1.5 mm).

Figure 12. *Oecetis* species A, larva and case. (A) head, dorsal; (B) pronotum, dorsal (scale bar = 0.5 mm); (C) head, lateral (scale bar = 0.5 mm); (D) ventral apotome; (E) case (scale bar = 1.5 mm).

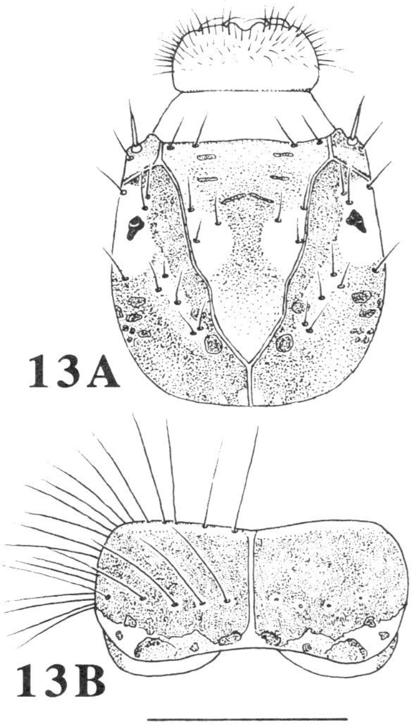

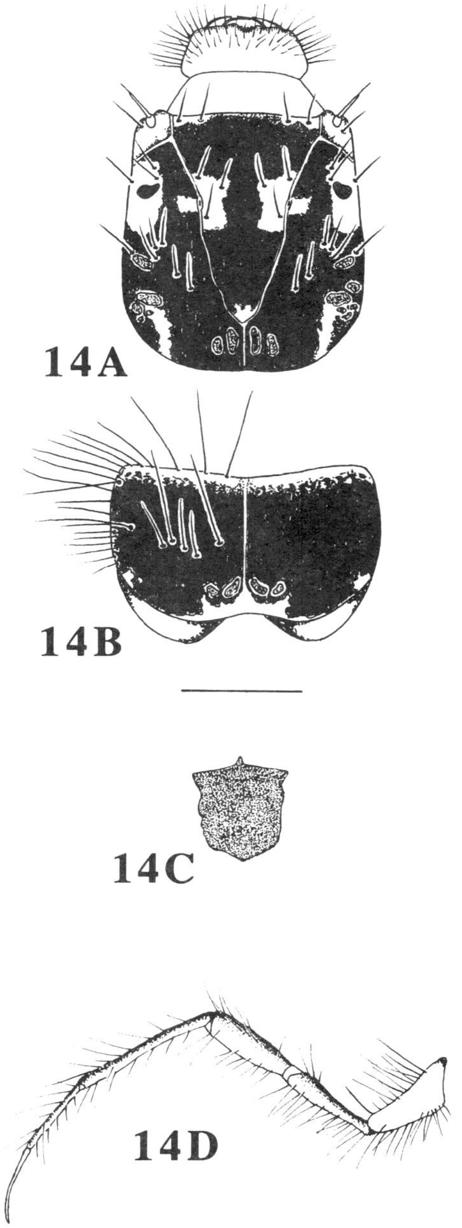

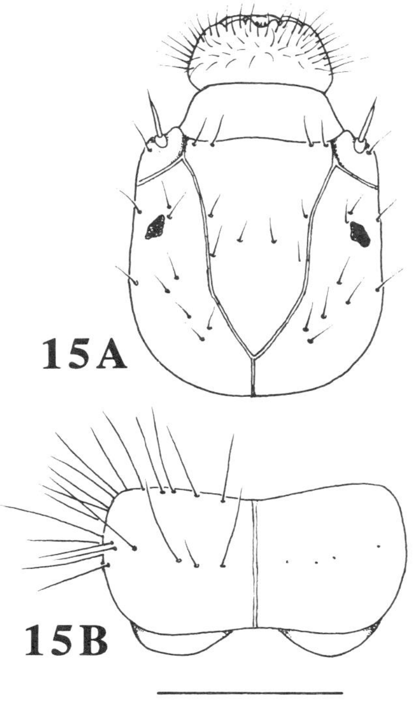

Figure 13. *Oecetis* species B, larva. (A) head, dorsal; (B) pronotum, dorsal (scale bar = 0.5 mm).

Figure 14. *Oecetis* species C, larva. (A) head, dorsal; (B) pronotum, dorsal (scale bar = 0.5 mm); (C) ventral apotome;(D) left hindleg, lateral.

Figure 15. *Oecetis* species D, larva. (A) head, dorsal; (B) pronotum, dorsal (scale bar = 0.5 mm).

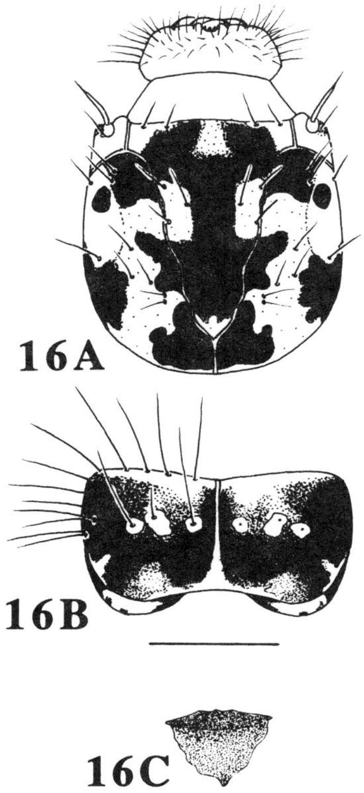

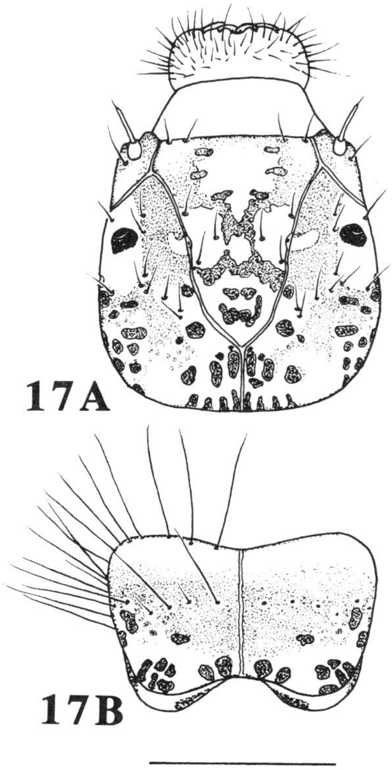

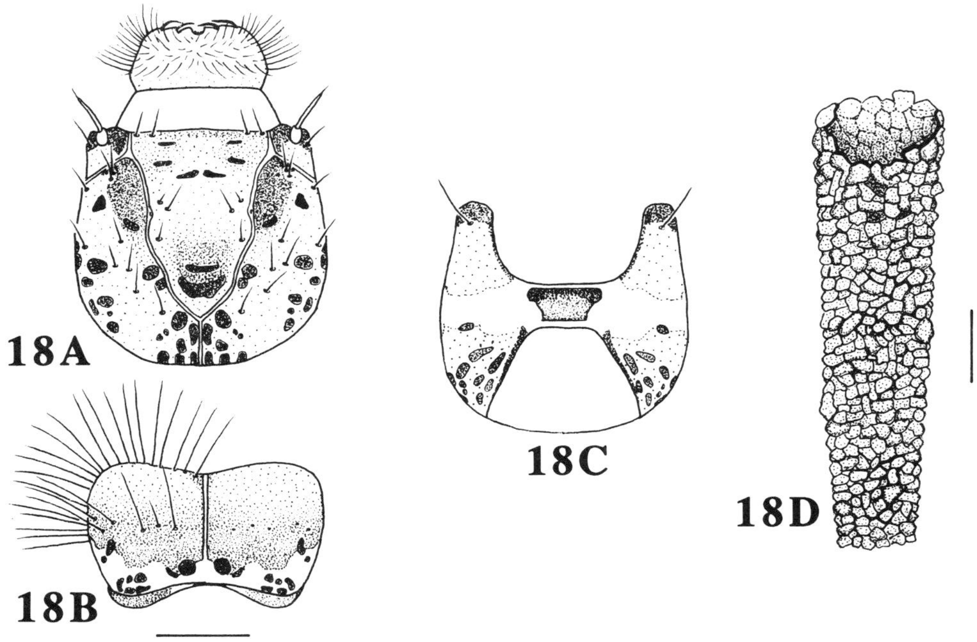

Figure 16. *Oecetis* species E, larva. (a) head, dorsal; (b) pronotum, dorsal (scale bar = 0.5 mm); (c) ventral apotome.

Figure 17. *Oecetis* species F, larva. (A) head, dorsal; (B) pronotum, dorsal (scale bar = 0.5 mm).

Figure 18. *Oecetis ochracea*, larva and case. (A) head, dorsal; (B) pronotum, dorsal (scale bar = 0.5 mm); (C) head, ventral; (D) case (scale bar = 1.5 mm).

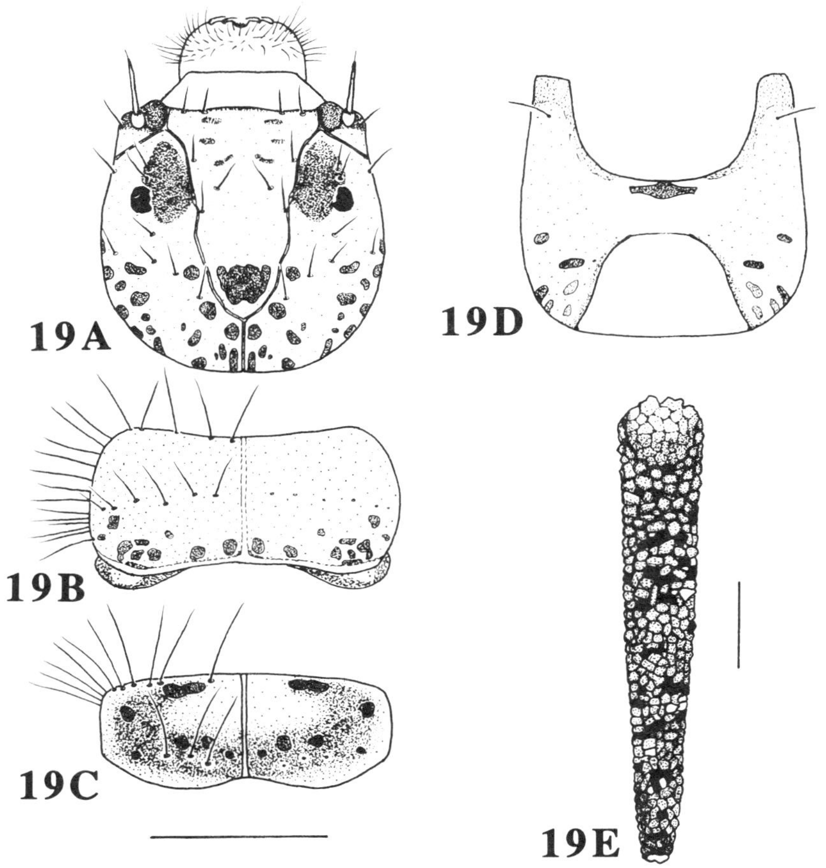

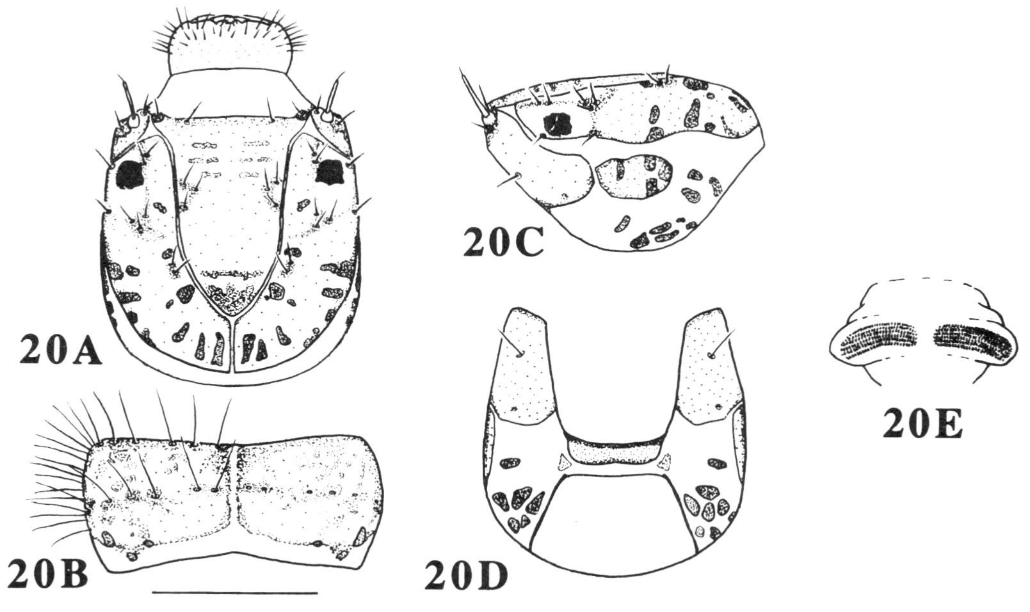

Figure 19. *Oecetis immobilis*, larva and case. (A) head, dorsal; (B) pronotum, dorsal; (C) mesonotum, dorsal (scale bar = 0.5 mm); (D) head, ventral; (E) case (scale bar = 1.5 mm).

Figure 20. *Oecetis avara*, larva and case. (A) head, dorsal; (B) pronotum, dorsal (scale bar = 0.5 mm); (C) head, left lateral; (D) head, ventral; (E) dorsal hump of abdominal segment I, dorsal.

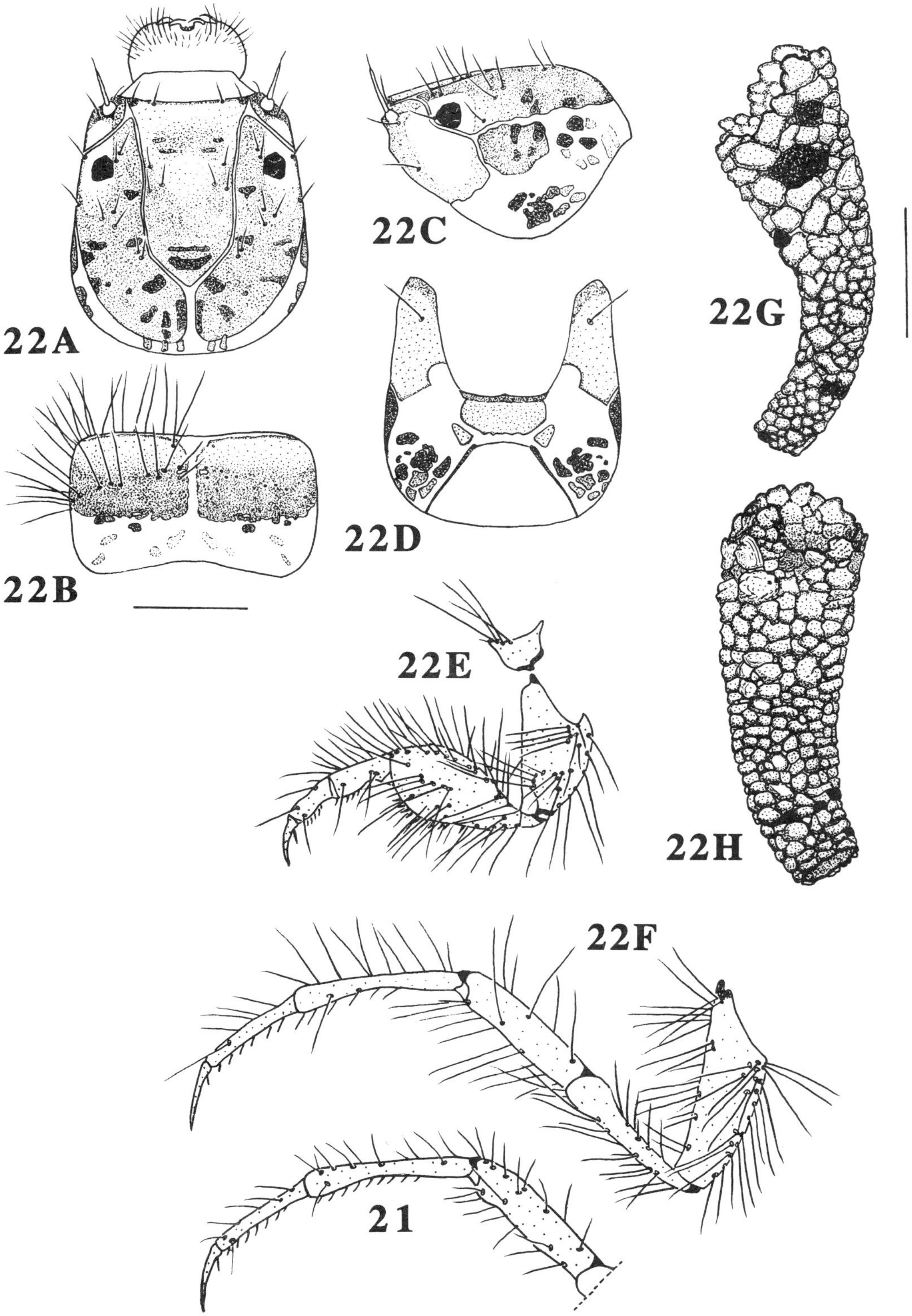

Figure 21. *Oecetis avara*, left hindleg, lateral.

Figure 22. *Oecetis disjuncta*, larva. (A) head, dorsal; (B) pronotum, dorsal (scale bar = 0.5 mm); (C) head, left lateral; (D) head, ventral; (E) left foreleg, lateral; (F) left hindleg, lateral; (G) case, third instar (scale bar = 1.5 mm); (H) case, final instar.

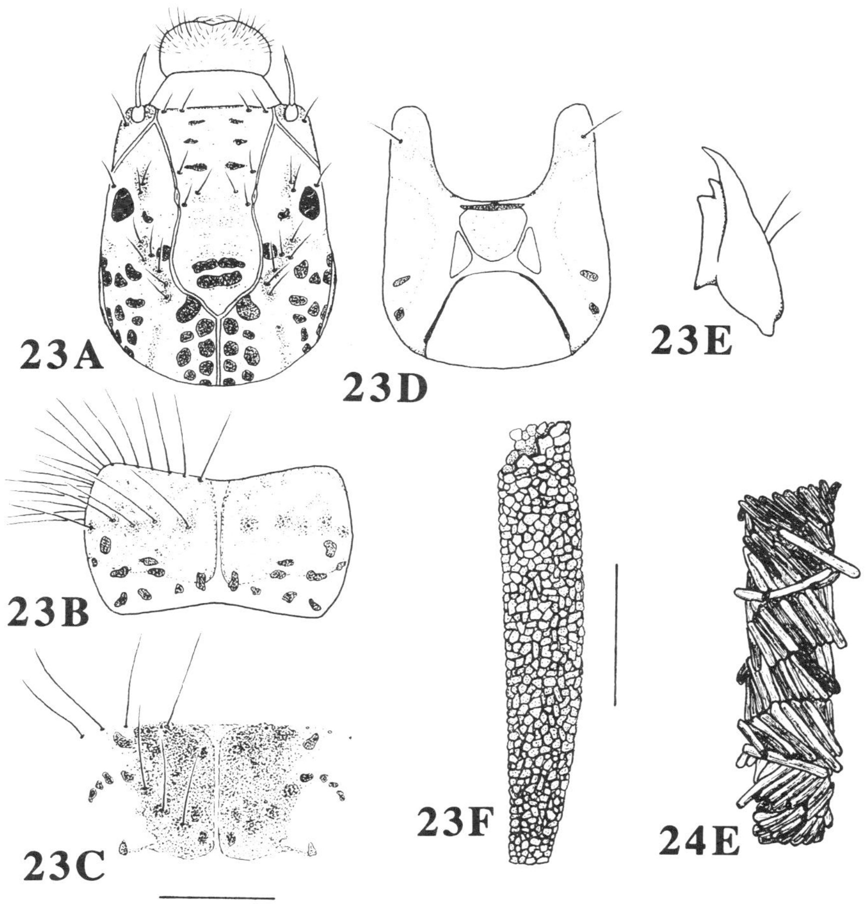

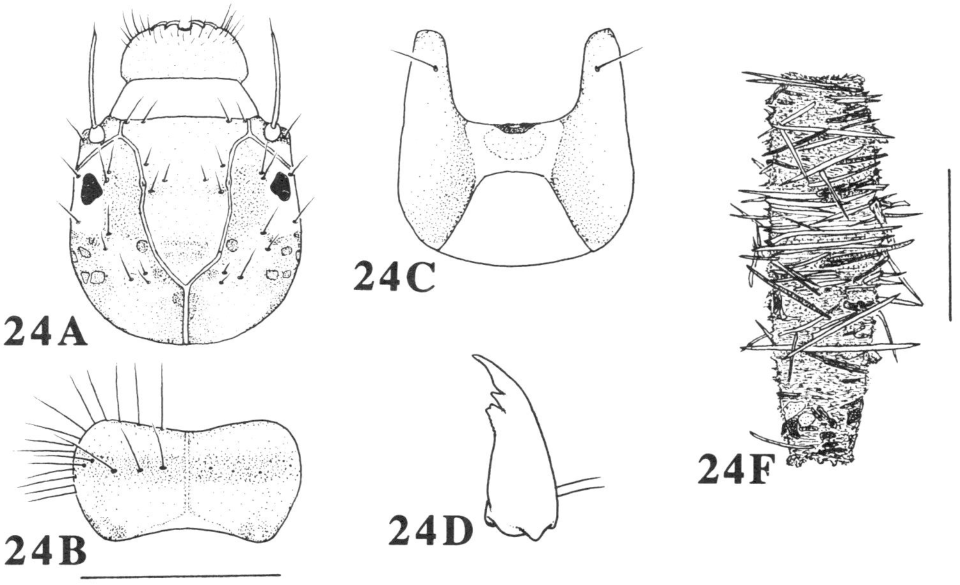

Figure 23. *Oecetis osteni*, larva and case. (A) head, dorsal; (B) pronotum, dorsal (scale bar = 0.25 mm); (C) mesonotum, dorsal; (D) head, ventral; (E) left mandible, ventral; (F) case (scale bar = 0.5 mm).

Figure 24. *Oecetis parva*, larva and cases. (A) head, dorsal; (B) pronotum, dorsal (scale bar = 0.25 mm); (D) head, ventral; (E) left mandible, ventral; (E-F) cases (scale bar = 2.0 mm).

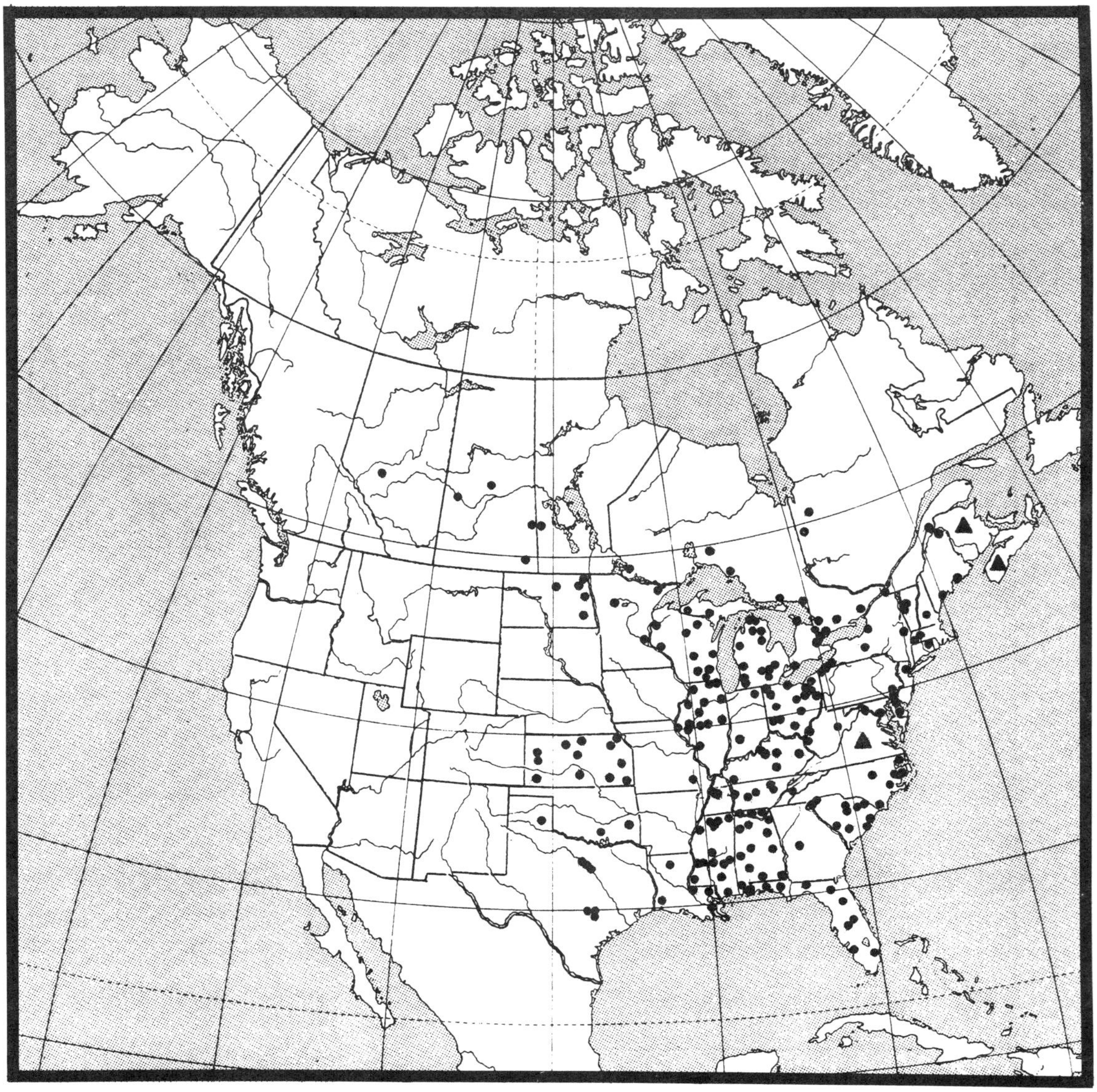

Map 1. Distribution of *Oecetis cinerascens* in North America. Circles represent actual collection localities, triangles indicate literature records without specific locality information.

Map 2. Distribution of *Oecetis georgia* in North America.

Map 3. Distribution of *Oecetis persimilis* in North America. Circles represent actual collection localities, triangles indicate literature records without specific locality information.

Map 4. Distribution of *Oecetis eddlestoni* in North America. Circles represent actual collection localities, triangles indicate literature records without specific locality information.

Map 5. Distribution of *Oecetis morsei* in North America.

Map 6. Distribution of *Oecetis ouachita* in North America.

Map 7. Distribution of *Oecetis ozarkiensis* in North America.

Map 8. Distribution of *Oecetis scala* in North America. Circles represent actual collection localities, triangles indicate literature records without specific locality information.

Map 9. Distribution of *Oecetis sphyra* in North America.

Map 10. Distribution of *Oecetis arizonica* in North America.

Map 11. Distribution of *Oecetis ditissa* in North America. Circles represent actual collection localities, triangles indicate literature records without specific locality information.

Map 12. Distribution of *Oecetis nocturna* in North America. Circles represent actual collection localities, triangles indicate literature records without specific locality information.

Map 13. Distribution of *Oecetis porteri* in North America.

Map 14. Distribution of *Oecetis pratelia* in North America.

Map 15. Distribution of *Oecetis* species A in North America.

Map 16. Distribution of *Oecetis* species B in North America.

Map 17. Distribution of *Oecetis* species C in North America.

Map 18. Distribution of *Oecetis* species D in North America.

Map 19. Distribution of *Oecetis* species E in North America.

Map 20. Distribution of *Oecetis* species F in North America.

Map 21. Distribution of *Oecetis* species G in North America.

Map 22. Distribution of *Oecetis ochracea* in North America. Circles represent actual collection localities, triangles indicate literature records without specific locality information.

Map 23. Distribution of *Oecetis immobilis* in North America. Circles represent actual collection localities, triangles indicate literature records without specific locality information.

Map 24. Distribution of *Oecetis avara* in North America. Circles represent actual collection localities, triangles indicate literature records without specific locality information.

Map 25. Distribution of *Oecetis disjuncta* in North America.

Map 26. Distribution of *Oecetis osteni* in North America. Circles represent actual collection localities, triangles indicate literature records without specific locality information.

Map 27. Distribution of *Oecetis daytona* in North America.

Map 28. Distribution of *Oecetis parva* in North America.

PUBLICATION POLICY OF THE OHIO BIOLOGICAL SURVEY

Any manuscript to be considered by the Ohio Biological Survey must first and foremost be scientifically credible and accurate. The Survey is most interested in manuscripts which concentrate on Ohio's biota. A monograph outlining the status of a particular taxon within Ohio, an in-depth study of a particular Ohio natural area, or an amalgamation of information of use to Ohio's biologists are examples of areas of paramount interest to the Survey.

Secondly, the Ohio Biological Survey will consider a manuscript dealing with a topic in which Ohio, or an Ohio natural area, or an Ohio native taxon is an integral part. Any study encompassing the Ohio River drainage basin, the Great Lakes region, or eastern North America, for example, would be given a reading by the Ohio Biological Survey.

Finally, the Ohio Biological Survey will consider manuscripts by Ohio authors. It is preferable that these manuscripts meet either of the criteria above, and such manuscripts will be given priority. However, if an Ohio author creates a manuscript meeting the criteria for one of the Survey's publication categories, it will be read with an eye to publication.

It should be remembered, however, that any manuscript, regardless of origin or topic, must be first approved by the Survey's Editorial Committee, and then pass peer-review by several experts in the field of study.

Ohio Biological Survey
1315 Kinnear Road
Columbus, OH 43212-1192

PUBLICATIONS CRITERIA

I. Ohio Biological Survey *Bulletin New Series*

1. Original work
2. Data-rich, and therefore probably lengthy
3. Comprehensive
4. Having potential archival value
5. Historical perspective on the problem plus the rationale for the current investigation
6. "Literature Cited" must be included
7. If a taxonomic work, synonomies and diagnoses should be included and new taxa may be raised; checklists, distribution data, and keys with accompanying figures should be included
8. Bibliography, Foreword, Preface, Ecological/Behavioral Information—these sections are optional
9. Manuscript will be peer-reviewed by two or more experts in the field

II. Ohio Biological Survey *Miscellaneous Contribution*

1. Original work
2. Of variable length but usually shorter than a *Bulletin*
3. The category *Miscellaneous Contributions* may encompass proceedings of meetings, annotated bibliographies, anthologies, catalogs, atlases, and others
4. Where appropriate, *Miscellaneous Contributions* will have a "Literature Cited" section, or perhaps more than one such section
5. Manuscripts will be peer-reviewed by two or more experts in the field

III. Ohio Biological Survey *Notes*

1. Original work
2. Focused on a precisely defined topic
3. Of variable length, but concise organization
4. *Notes* will present the results of biological surveys, unannotated bibliographic compilations, unified short biological studies, and/or short natural historical offerings; several short papers of a theme may be combined
5. Where appropriate, *Notes* will have a "Literature Cited" section, or perhaps more than one such section
6. Manuscript will be peer-reviewed by two or more experts in the field

IV. Ohio Biological Survey *Informative Publications*

1. Quickly-produced and disseminated amalgamations of information useful (primarily) to Ohio biologists (lists; bibliographies; meetings' abstracts)
2. Short; likely to be the last word on a subject for only a brief time
3. Internal review only

REQUIREMENTS FOR AUTHORS

Please refer first to the Publications Criteria listed previously.

I. Sections, and suggested order (please bear in mind that not all sections need be included, and certain sections are neither necessary or even desirable for all four publication categories).

A. Abstract
B. Acknowledgements
C. Lists of (Figures, Tables, Maps, Photos, Plates)
D. Introductory Materials (could include behavior, ecology, geology, history—these could be subsections)
E Materials and Methods
F. Checklist
G. Results (could be set up as subsections)
H. Discussion (again, would probably be set up in subsections, especially for the *Bulletins New Series*, *Miscellaneous Contributions*, and *Notes*, and could include data analyses, diagnoses, population information, new taxa and the rationale for their creation)
I. Taxonomic Key(s)
J. Summary
K. Glossary
L. Literature Cited
M. Bibliography
N. Appendices
O. Indices

Prose style should be that of previously published *Bulletins New Series*, *Miscellaneous Contributions, Biological Notes*, or *Informative Circulars*. Spell out numbers one through ten, and use Arabic numerals for 11 and above. Use the active voice. Scientific names and the titles of books and journals will be italicized in the published work (see II. below). Abbreviations and symbols used should be used following the guidelines and lists given in the CBE Style Manual, 5th Edition. Use the International System of Units in all cases.

NEVER use footnotes.

II. Manuscripts should be submitted on computer disk as a Word Perfect for Windows file (the most recent version is preferred). The format of earlier Ohio Biological Survey publications should be followed. Scientific names and the titles of books and journals should be italicized. Tables should be placed in Word Perfect table format. If Pagemaker or QuarkExpress for Windows is used, then black and white artwork can be scanned right into place, as can black and white photographs. However, the original graphics and photographs must accompany the manuscript file.

III. Black and white photographs should be submitted as prints. Black and white line drawings should be submitted as PMTs (PMT is an abbreviation of "photomechanical transfer," and refers to the process by which a black and white piece of artwork is reduced or enlarged photographically, and then transferred mechanically in between two pieces of receiver paper and treated with an activator fluid, so that you end up with a positive representation of the drawing you want at the size you want. They are sometimes called "stats," as well). Color photographs should be submitted as transparencies — slides or photographs up to 8.5" x 11."